NLP Application

Applying Natural Language Processing (NLP) concepts to help humans in their daily life, this book discusses an automatic translation of an unstructured Natural Language Question (NLQ) into a Structured Query Language (SQL) statement. Using SQL as a Relational DataBase (RDB) interaction language, database administrators or general users with little to no SQL querying abilities are provided with all the knowledge necessary to perform queries on RDBs in an interactive manner.

Key Features:

- Includes extensive and illustrative examples to simplify the discussed concepts
- Discusses a novel, and yet simple, approach to NLP
- Introduces a lightweight NLQ into SQL translation approach through the use of RDB MetaTables as a Hash table
- Extensive literature review and thorough background information on every tool, concept and technique applied

Providing a unique approach to NLQ into SQL translation, as well as comprising disparate resources on NLP as a whole, this shortform book is of direct use to administrators and general users of databases.

Ftoon Kedwan is a recognized expert and researcher in the field of Artificial Intelligence, Software Engineering, Big Data Analysis, Machine Learning, and Health Informatics. Dr. Kedwan is experienced in industry and academics and is affiliated with many recognized organizations and universities, such as Queen's University, the Royal Military College of Canada (RMC), St. Francis Xavier University (StFX), the University of Prince Mugrin (UPM), and the Saudi National Guard Hospital.

NLP Application
Natural Language Questions and SQL using Computational Linguistics

Ftoon Kedwan

CRC Press
Taylor & Francis Group
Boca Raton London New York

CRC Press is an imprint of the
Taylor & Francis Group, an **informa** business

Cover Image Credit: Shutterstock.com

First edition published 2024
by CRC Press
2385 NW Executive Center Drive, Suite 320, Boca Raton, FL 33431

and by CRC Press
4 Park Square, Milton Park, Abingdon, Oxon, OX14 4RN

CRC Press is an imprint of Taylor & Francis Group, LLC

© 2024 Ftoon Kedwan

ISBN: 978-1-032-53835-8 (hbk)
ISBN: 978-1-032-53837-2 (pbk)
ISBN: 978-1-003-41389-9 (ebk)

DOI: 10.1201/b23367

Typeset in Times
by SPi Technologies India Pvt Ltd (Straive)

Contents

Preface

This book applies Natural Language Processing (NLP) concepts to help humans in their daily life. It discusses an automatic translation of an unstructured Natural Language Question (NLQ) into a Structured Query Language (SQL) statement. SQL is used as a Relational DataBase (RDB) interaction language with special query syntax and a computer-executable artificial language. This way, DataBase (DB) administrators or general users with little or no SQL querying abilities can perform queries on RDBs in an interactive manner. The Human–Computer Interaction (HCI) happens using users' NLQs, which is in English in the proposed research. Users do not need to know any RDB schema elements or structures such as tables' names, relationships, formats, attributes, or data types. The RDB schema is a brief description of the RDB elements' organization, excluding any RDB values. In this work, a lightweight NLQ into SQL translation approach is implemented by utilizing an RDB MetaTable as a Hash table. The main goal is to exploit a manually written rule-based mapping constraints algorithm. This algorithm maps NLQ tokens' semantic/syntactic information into RDB elements' semantic roles (i.e., value, attribute) in addition to the Wh-Words (e.g., What, Where, How, and Who) of the actual data and the relationships between the attributes. via pairing and matching means. The matching RDB elements, called "identified lexica", are then mapped into the SQL clauses consistently for SQL generation and execution. The matching process uses a computational linguistic analysis mapping algorithm, represented in the MetaTables. This mapping algorithm proved to be efficient especially with small RDBs with an accuracy of 95% and is about 93% accurate with larger RDBS.

Introduction

1

NLP is a subfield of computer science and engineering under the field of Artificial Intelligence (AI), as illustrated in Figure 1.

It is developed from the study of language and computational linguistics [1, 2] and often used to interpret an input Natural Language Question (NLQ) [3]. NLP's goal is to analyze and facilitate the interaction between human language and computing machines. HCI becomes a part of NLP when the interaction involves the use of natural language. Under NLP, there are several subareas, including Question Answering Systems (QAS) [4], such as Siri for iPhones, [5] and summarization tools [6]. Such tools produce a summary of a long document's contents or even generate slide presentations. Machine real-time translation [7], such as Google Translate [8] or BabelFish [9], are among other examples of NLP subareas. In addition, document classification [10] via learning models is a famous NLP subarea. It is used to train the classification algorithm to identify the category a document should be placed under, for example, news articles or spam filtering classification. Speech Recognition Models [11] are another NLP subarea that recognizes spoken language words which work best only in specific domains.

In the current research, the framework starts with a processing of simple Online Transactional Processing (OLTP) type of queries. OLTP queries are simple SELECT, FROM and WHERE statements of the Structured Query Language (SQL), which is the simplest query form. As in Figure 1, NLP uses deep learning techniques as part of the AI area. It focusses on computational linguistics to analyze HCI in terms of the language used for this interactive communication. Basically, NLP bridges the gap between computers and humans and facilitates information exchange and retrieval from an adopted

FIGURE 1 Research area breakdown.

DOI: 10.1201/b23367-1

DataBase (DB), which is, in this case, a Natural Language Interface for DataBase (NLIDB).

A Relational DataBase (RDB) model [12], which was originally introduced in 1970, is used as the basis of a background data storage and management structure. RDB has been chosen as a data storage media for the proposed research because of the relationships between its entities, including their table attributes and subfields and their values. These relationships hold significant information themselves as if they were whole separate entities. In addition, the information stored on those relationships proved to increase the accuracy of data retrieval as will be demonstrated later in Chapter 6 on implementation testing and performance measurements.

RDB's elements (i.e., Table, Attribute, Relationship, etc.) representation in Figure 2 describes the relationships between the entity sets to express parts of Post-Traumatic Stress Disorder (PTSD) RDB semantics. Therefore, an Entity-Relational Diagram (ERD) [13] was used to demonstrate the RDB structure because this makes data relationships more visible as follows:

- An entity sets represent a table, i.e., "Patient" table.
- The entity's features represent specific table's attributes, i.e., "P_Name".
- Any instance of a specific attribute represents an attribute's value, i.e., "Sarah".
- The directed relationships represent entities' act or impact on other entities, i.e., "take".

In 1976, Chen [13] was the first to graphically model RDB schema entities using ERD to represent NLQ constructs. In [14], ERD was used to represent NLQ constructs by analyzing the NLQ constructs' inter-relationship with the ERD or even with the Class Diagram Conceptual Schema [15]. However, NLQ

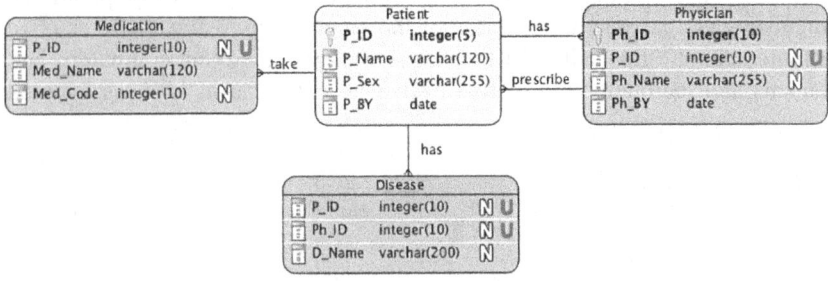

FIGURE 2   ERD for Post-Traumatic Stress Disorder (PTSD) RDB.

constructs intra-relationships were not studied in the previous works in terms of mapping the constructs into an RDB query language, such as SQL, which will be implemented in the current research work. Intra-relationships are the lexical dependencies between the sentence constructs (i.e., words), such as the relationship between the verb and its object. This type of relationship supports the mapping of tokens into lexica.

Furthermore, RDB enables NLQ into SQL mapping using RDB schema MetaTables, such as in Tables 1 and 2 (see Chapter 2). MetaTables are data repositories that act as data dictionaries which describe the RDB elements and the relationships between them using data annotations, span tags and synonyms attachment. In addition to the used RDB MetaTables, the NLQ context and the situation-based linking of knowledge stored in multiple connected RDB tables all help enhance the accuracy of data retrieval. Under the scope of the proposed research, the major focus will be on finding an answer to the question: How can we translate an unstructured full-text NLQ expression to an SQL statement using RDB schema MetaTables such that it produces accurate results? This translation mechanism and its framework design, starting from the NLQ interface and up to identifying the equivalent SQL clauses, aims to achieve other secondary tasks, such as:

Abbreviation support; so that if a user asks about a patient's weight using the abbreviation "kg", it is recognized as a "kilogram".

- Support for SQL's syntactic constructs' (keywords such as SELECT, FROM, etc.) synonyms or the absence of them in the NLQ.
- Multiple columns SELECT; to recognize multiple NLQ's main nouns or noun phrases.
- Converting operators and numbers (i.e., Equal, Three) into numerical forms and symbols (i.e., =, 3).
- Deriving tables' or attributes' names from the literal values mentioned in the NLQ.
- Support for propositional terms (i.e., above, below, between).
- Considering all conditional terms in the NLQ and converting them into multiple WHERE conditions. This shall apply whether all conditions are applied on the same or different RDB elements.
- Support for aggregate functions (e.g., convert 'highest' to 'MAX (Hight)' or 'youngest' to 'MIN(Age)').

Table 1 is an example of the NLQ MetaTable that breaks down the entered NLQ into its subsequent tokens. Table 2 is an example of the RDB elements MetaTable that explains each RDB element in terms of its nature, category, syntactic role etc. These two tables will be referenced to and elaborated on frequently throughout this research document.

Machine-readable instructions are mandatory to access any type of DB. This emphasizes the need to find a mapping mechanism between NLQ and the RDB query languages, such as SQL. Common semantics between NLQ and artificial languages can be discovered by analyzing the language semantic and syntax roles. This research work solves the NLQ into SQL translation problem by manually writing a rule-based mapping algorithm at the word-processing level for automatic mapping. The manual work is on creating the rule-based algorithm. After the algorithm is developed, the mapping process shall be automatic.

The aim of this work is to maintain a simple algorithmic configuration with high outcome performance. Avoiding the reliance on a huge annotated corpus or written patterns of translation examples is also very important, except in the case of simple algorithmic rules. This research idea overcomes any poor underlying linguistic tools' performance such as named entity tagger, tokenizer, or dependency parser. This research contribution is a rule-based algorithm that applies a mapping association between the NLQ's semantic/syntactic information, the SQL's syntactic information and the RDB elements' semantic roles and it offers an effective translation mechanism to convert NLQs into SQLs.

BASIC RESEARCH FRAMEWORK ORGANIZATION

In the implementation of the proposed algorithm, and as illustrated in Figure 3, user's NLQ is accepted as an input into a given NLIDB, together with its corresponding RDB schema. Next, an NLQ into SQL translation is performed by an underlying multi-layered NLP framework. Afterwards, the system can basically understand the NLQ and respond to it with its equivalent SQL query.

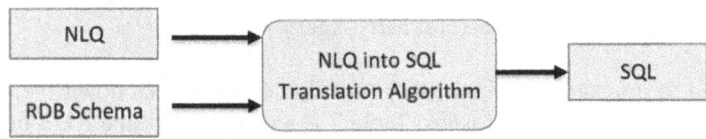

FIGURE 3   Basic research organization diagram.

In the following sections, Chapter 2 is a background study on the research work pipeline components, which acts as an introduction to the actual research work implementation plan described in Chapter 4. Chapter 3 is a literature review of the main research contributions, which are the two mapping algorithms. Chapter 5 is a running example using an implementation user case scenario to illustrate the role of each component. This chapter also presents a complete summarization of the algorithm's processes. Chapter 6 sets out the algorithm's implementation testing and validation methods, in addition to the implementation performance measurements used to illustrate the success factors of the proposed research work. Chapter 7 presents the framework implementation results and discussion and then, finally, Chapter 8 presents the research conclusion and outlines the scope for future study.

Background Study

2

For terminology clarification purposes, Figure 4 explains the Natural Language Query (NLQ) words journey to tokens and then to lexica throughout the processes executed and summarized on the arrows.

NLQ INPUT PROCESSING INTERFACE, THE NLIDB

This first step is a point of interaction with the user. The user interface could be either a simple coded data input interface or a web-based Graphical User Interface (GUI). GUIs use web design languages such as HTML, CSS, and PHP. This interface is used to enter a question in an NLQ form. Generally, NLIDB could be any of the following options:

Interactive Form-Based Interface [16]

Though this looks attractive, it does not always retrieve data variations from the DB as we can easily define in a formal SQL statement.

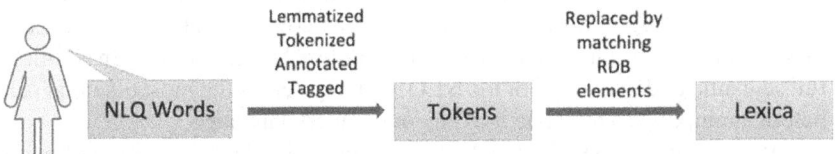

FIGURE 4 NLQ words, tokens and lexica terminology clarification.

DOI: 10.1201/b23367-2

Keyword-Based Query Interface [17]

Extracts keywords from the input NLQ and identifies whether those keywords are domain-specific or language constructs. Next, the system applies translation rules from the generated knowledge base to use those keywords to form an SQL query.

NLQ-Based Interface [18]

Unlike previous options, this approach tries to actually understand the NLQ first before converting it into an SQL statement. To do this, the system applies lexicon breakdown and NLP syntactic and semantic analysis on the input NLQ. Lexica is the plural of lexicon, which is the name of a matching RDB element with a stemmed NLQ token.

The NLQ questions are either:

1) Simple straightforward questions with simple linguistic structures that have one question word, complete semantic meanings, and no ambiguous expressions; or

2) Complicated general questions with lots of ambiguous expressions and fuzzy predicates. The ambiguity could be due to omitting necessary words, adding unrelated words, including more than 1 question word, or the lack thereof.

Furthermore, NLQ questions can be further categorized as questions asking for a specific entity, quantity, rank of a list of entities or a proportion ratio. They can also be interrogative questions, being either imperative or declarative in a negative or affirmative form. The NLP linguistic tools are trained to capture those question structures and types.

In the proposed research, an NLQ-based interface is chosen as an interface to the Natural Language Interface to DataBase (NLIDB) system. This is because it is the most natural way of communicating with a computing machine. It is also the most desirable way of communication for decision-makers, who are the target audience for the proposed research work. In addition, full semantic understanding of the NLQ by the system is the most essential part before attempting to translate it to any other query language.

With regard to the adopted NLQ type, only simple straightforward questions, which is the first NLQ question type as explained earlier, are considered. This is because the main focus and contribution of the proposed research work is the mapping mechanism, not resolving or disambiguating misstructured or incomplete NLQs. Yet there is a simple disambiguation module implemented for simple NLQ problems. However, addressing the complicated general questions (i.e., with aggregation or negation) is one of the main suggestion for future study.

PART OF SPEECH (POS) RECOGNITION

To recognize part of speech (POS) for NLQs, the following subsequent layers must be clarified first.

Linguistic Components Layers

In the proposed system, morphological analysis is performed during the lexical analysis in order to identify keywords' properties and break them down into their basic components. A good example for morphological and lexical analysis is WordNet [19]. This traces connections between keywords and separates the words from their suffixes or affixes (e.g., Client's to 'client' and 's'). In addition, WordNet separates words from their punctuations (a.k.a. non-word tokens).

Other linguistic components, which are also part of the NLP tools in the current research framework, include the tokenizer, the POS tagger, the dependency tree parser, Named Entity Recognition (NER), and syntactic and semantic analyzers such as Link Grammar Parser [20] and Stanford Parser [21]. Those NLP tools are principally used for transforming NLQ tokens into a parse tree structure to then map them into their appropriate SQL query template. Discourse Analysis [22] and Pragmatics [23] also fall under the latter NLQ translation method. SQL syntactic parse trees are usually 100% accurate. The NLQ parse trees are also accurate but usually introduce some noise affecting the overall NLQ features representation accuracy [24].

The NLQ grammar is distributed to various linguistic areas, including discourse, pragmatics and text theory [23]. More specifically, the included linguistic areas are Cognitive Designs [25], Descriptive Convenience Modules [26] or specific techniques such as syntax, semantics, morphology, phonology and lexica.

Syntactic parser (rule-based)

Syntax is the study of how words are combined to form sentences. It includes the analysis of POS to show how words are related to one another, and their aggregations to form larger sentence constructions. The syntax analyser ignores sentences that do not conform to a proper language grammar or rules of how words are supposed to be combined.

POS tags are needed before sentence parsing. Syntactic parsing analyzes complex sentences with large amounts of ambiguousness and conditional relationships. It also defines the scope of exclusion and probability statements (statements on the probability of inclusion [27]).

Under syntactic parsing, there are subtasks that also take place, including:

i.   Text Structure Analysis: This is the study of how similarly comparable words are and how their textual types are constructed to compose larger textual constructions [28].

ii. Lemmatizer: This module derives the original form (the root word) of each NLQ word, which happens before tokenization.

iii. Tokenizer: Entered NLQ is tokenized by separating the NLQ textual words into separate tokens. Those tokens will be stored and passed on to the next module for further analysis.

iv. POS Tagger: This is otherwise known as word category disambiguation or grammatical tagging. POS tagging takes into consideration both the word's definition and its context in the sentence; to mark the word in the text or corpus to belong to a particular POS.

Semantic parser (rule-based)

Semantics is the study of the actual meaning of a word such that they relate a certain language word with its intentional meaning representation. An example of semantics analysis is relating a syntactic constituent with a predicate [29]. This is because all structures created by the syntactic analyzer have meanings. In the case when there was no equivalent meaning representation to a certain syntactic structure, the whole sentence is rejected and ignored by the semantic parser considering it semantically anomalous.

NLQs are processed based on their accurate semantic properties, indicated by underlying NLP linguistic tools such as the semantic parser. It is important to define the context in an NLQ with information further than the equivalent lexical properties. As such, verbs, subjects, objects and their relationships must be identified.

Under the semantics analysis and parsing, there are the following definitions:

i.   Pragmatics [23]: is studying how the textual context affects expressions' meanings. It explores expression's implicit meaning using the sentence's current structure representation. Next, the expression is reinterpreted to determine its specific implicit meaning.

ii. Morphology [30]: is studying the units of meanings or functions (called morphemes) in a certain language. A morpheme can be a word, affix, prefix, or another word structure.

iii. Phonology [31]: is studying the sound patterns of a certain language to determine which phones are significant and meaningful (the phonemes). Phonology also studies syllables' structures and the needed features to describe and interpret the discrete units (segments) in sentences.

iv. Phonetics [32]: is studying the sounds of spoken language. Phonetics studies how a phoneme is made and perceived. A phoneme is the smallest unit of an individual speech sound.

Lexicon

Lexicon studies aim at studying a language vocabulary, that is words and phrases. In the current research work, RDB lexica are stored in the RDB MetaTable which is used to map the NLQ words to their formal representations in an RDB (i.e., table names, attribute names, etc.). Lexica are analyzed by both syntactic and semantic parsers.

Intermediate Language Representation Layer

Those layers use a knowledge base to assist the NLQ with the SQL translation process. A good example for this is the discourse representation structure. This type of structure converts NLQs into SQLs using an ontology-based semantic interpreter as an intermediate language representation layer. Discourse Integration Analysis [22] is the study of information exchange such as in conversations where sentences meanings may change according to the preceding/proceeding sentences. An example is the use of pronouns in sentences such as "He" which will depend on the subject or the actor in the preceding sentence. In order for the discourse analysis process to be thorough, the NLQ tokens must be annotated. Hence, the annotator is essential in this method.

Annotator

Annotations are metadata that provide additional information on particular data. Natural language datasets are called corpora (the plural of corpus). When a corpus is annotated, this is called an annotated corpus. Corpora are used to train Machine Learning Algorithms (MLA). The annotation types include:

• POS Annotation.
• Phrase Structure Annotation.
• Dependency Structure Annotation.

In the current research work, all of the above underlying NLP layers are considered and implemented as part of the POS recognition module, with the exception of the intermediate language representation layer. The employed POS recognition Python library is "speech_recognition". This module is essential to the thorough understanding of the given NLQ for proper and accurate translation.

DISAMBIGUATION

In the current research work, disambiguation is only required when the mapper finds more than one matching RDB element to a certain NLQ token. Generally, a word's meaning disambiguation has special techniques and algorithms. An example is the statistical keyword meaning disambiguation process using N-Gram Vectors [33], where N ranges from 1 to the length of the text. N-Gram Vectors' statistics are gathered using a training corpus of English language, while a customised corpus gather statistics as the system is being used. The latter corpus requires that the user reviews the presented NLQ interpretation and makes any necessary changes before submitting it for execution. N-Gram Vectors are principally used for capturing lexical context. It is considered as a measure of how likely a given token has a particular meaning in a particular user NLQ input. This is because every token's meaning depends on the context in which it was used. This meaning comparison procedure follows a vector similarity ranking measure where the higher the meaning rank the closer it is to the true meaning.

Other words' meaning disambiguation studies involve computational linguistics and statistical language understanding. Such studies focus on word sense disambiguation within a large corpus. Another method to disambiguate a word sense is using word collocations. The word collocations method measures the meaning likelihood of more than one word when they all exist in the same NLQ. Iftikhar [34] proposed a solution to the disambiguation problem by parsing domain-specific English NLQs and generate SQL queries by using the Stanford Parser [21]. This approach is widely used with the Not Only Structured Query Language (NoSQL) DBs for automatic query and design.

Rajender [35] used a controlled NLIDB interface and recommended SQL query features to reduce the ambiguity in an NLQ input. This is because the less ambiguous the domain the more accurate results the system can produce. What can also help resolve such ambiguity is the weighted relationships or links [36] between RDB elements. In this method, the relationship's weight increases by one each time that particular relationship is used. As such, it is a given that the bigger the relationship weight is the more often the related RDB elements are queried. This helps the NLIDB system recommend smarter options for the user to select from with ordered options, where the topmost option is the most likely.

In the current research work, NLQ disambiguation is not the main focus. Thus, a simple NLQ disambiguation module is used by applying Stanford CoreNLP [21] and "nltk.corpus" Python library [37]. Those tools are solely used to check for NLQ validity. A syntactic rules checker is also used for any NLQ grammatical mistakes. The adopted procedure is interactive where an error message pops up to the user, asking them to rephrase a word or choose from a few potential spelling corrections. A Naïve Bayes Classifier [38] is implemented to simply classify user's response as positive or negative (i.e., Yes or No).

MATCHER/MAPPER

The Matcher/Mapper module is considered the most complicated part in NLP science [14]. Therefore, a keyword-based search has attracted many researchers as the simplest mapping approach since keywords are explicitly identified. Researchers used it to improve the information retrieval from DBs and solve (or avoid) the challenge of understanding NLQ tokens and mapping them into the underlying DB or schema model [39]. The mapper can be Entity-Attribute-Value (EAV) Mapper [40], Entity Relational (ER) Mapper [13], or eXtensible Markup Language (XML) Documents Mapper [41]. NLQs are translated into SQL queries if the used DB or scheme model is EAV or ER. NLQ can be translated to XML documents only if the used system employs a document-based data model in the underlying information system. More background information on this module is given in Chapter 3, the literature review.

In the current research work, the adopted mapper is the EAV mapper in addition to the RDB relationships. NLQ tokens are mapped into RDB lexica using the NLQ and RDB MetaTables, Tables 1 and 2, respectively. The matching lexica will then be mapped with the SQL clauses. This mapping uses the proposed rule-based algorithm that is based on the observational assumptions table discussed later in Table 4 (in Chapter 4).

Table 1 is an example of the NLQ MetaTable that breaks down the entered NLQ into its subsequent tokens. Table 2 is an example of the RDB elements MetaTable that explains each RDB element in terms of its nature, category, syntactic role etc. These two tables will be referenced to and elaborated on frequently throughout this research document.

TABLE 1   NLQ MetaTable

WORDS	SYNTACTIC ROLE	CATEGORY	SYNONYMS
Sarah	Instance	Value	Person, Patient, Physician
Has	Verb	Relationship	Own, Obtain, Have
Physician	Noun	Attribute	Doctor, Provider, Psychiatrist, Surgeon

TABLE 2   RDB elements MetaTable

WORDS	SYNTACTIC ROLE	CATEGORY	DATA TYPE	PK/FK	ENCLOSING SOURCE	SYNONYMS
Disease	Noun	Table	Word	No	RDB: PTSD	Illness, Sickness
Take	Verb	Relationship	Word	No	Tables: Patient->Medication	Acquire, Absorb, Get, Gain
P_ID	Noun	Attribute	Digits	Yes	Attributes: Patient, Physician, Medication, Disease	Patient, Identification, Number, Person

SQL TEMPLATE GENERATOR

This layer uses input from previous layers in conjunction with the available MetaTables' data to translate NLQ into SQL. The complexity of SQL queries can be incredibly high because some of them might require aggregation, nesting or negation with multiple tables selection and joining. Therefore, logical expressions concatenations might be required to solve potential NLQ complexity (e.g., negation conjunction, subordinates, superlatives, etc.). Basic WHERE conditions are combined with AND, OR and NOT operators to fall onto the nested WHERE clauses where a second SELECT clause is nested in the WHERE clause.

SQL templates and their complexity classifications determine which SQL category a particular NLQ falls under. The chosen SQL template's clauses are filled with the matching RDB lexica. Furthermore, different translation systems support different SQL templates. Hence, accepted and recognized templates must be identified, together with their likelihood ratio and occurrence frequency.

SQL Statements' Types:

- Data Query Language (DQL): SELECT
- Data Manipulation Language (DML): INSERT, UPDATE, DELETE
- Data Definition Language (DDL): CREATE, ALTER, DROP
- Data Control Language (DCL): GRANT, REVOKE
- Transaction Control Language (TCL): BEGIN TRAN, COMMIT TRAN, ROLLBACK
- Data Administration Commands (DAC): START AUDIT, STOP AUDIT

In the proposed research work, the focus is on the DQL, which has one command phrase (SELECT) in addition to other supplementary clauses (e.g., WHERE, AS, FROM). DQL, though it has one main command phrase, is the most used phrase among SQL's other phrases, especially when it comes into RDB operations. The SELECT keyword is used to send an inquiry to the RDB seeking a particular piece of information. This could be done via a command line prompt (i.e., Terminal) or through an Application Program Interface (API). This research proposes a translation algorithm from NLQ into SQL using the SELECT command phrase and its supplementary clauses.

SQL EXECUTION AND RESULT

At this stage, the end user must establish a connection with the RDB to send along generated SQLs. This layer performs all Database Management System (DBMS) functions. Once the translation process is done and the SQL statement is generated, the query is executed against the RDB. The query results are integrated in the form of raw data, that is columns and rows, as NLIDB system output.

The query results may get passed back to the first layer to produce a proper response to the end user in a natural language form. This is a reverse process to the original proposed research where SQL is the input and NLQ is the output for answer generation. This reverse process requires a discourse representation structure. The conversion from SQL into NLQ requires an ontology-based semantic interpreter. This interpreter is an intermediate language representation layer, which is not used in the original proposed research work.

Literature Review

3

RELATED WORKS

NLP

LUNAR [42–44] was developed in **1971** to take NLQs about a moon's rock sample and present answers from 2 RDBs using Woods' Procedural Semantics to reference literature and an Augmented Transition Network (ATN) parser for chemical data analysis. However, it only handled 78% of the NLQs due to linguistic limitations as it manages a very narrow and specific domain.

Philips Question Answering Machine (Philiqa) [4] was developed in **1977**. Philiqa separates the syntactic and semantics parsing of NLQ as the semantic parser is composed of three layers, namely English Formal Language, World Model Language, and schema DB metadata.

LIFER/LADDER [45] was developed a year later, in **1978**, as a DB NLP system interface to retrieve information regarding US Navy ships. It used a semantic grammar to parse NLQs. Although LIFER/LADDER supported querying distributed DBs, it could only retrieve data for queries related to 1 single table, or more than 1 table queries having easy join conditions.

In **1983**, ASK [46] was developed as a learning and information management system. ASK had the ability to communicate with several external DBs via the user's NLQ interface. ASK is considered a learning system due to its ability to learn new concepts and enhance its performance through user's interaction with the system.

TEAM [16] was developed in **1987** as an NLIDB with high portability and easy configurability on any DBA system without compatibility issues, which negatively affected TEAM's core functionality. NLIDB is an NL-based query interface which the user can use as a means of interaction with the DB to access or retrieve data using NLQs. This interface tries to understand the NLQ by parsing and tokenizing it to tokens or lexicons, and then applying syntactic or semantic analysis to identify terms used in the SQL query formation.

In **1997**, [17] a method of conceptual query statement filtration and processing interface using NLP was defined to essentially analyze predicates using full-fledged NL parsing to finally generate a structured query statement.

In **2002**, Clinical Data Analytics Language (CliniDAL) [27] initiated a solution for the mapping problem of keyword-based search using the similarity-based Top-k algorithm. Top-k algorithm searches for k-records of a dictionary with a significant similarity matches to a certain NLQ compared to a predefined similarity threshold. This algorithm was successful with accuracy of around 84%.

In **2002**, a Vietnamese NLIDB interface was developed for the economic survey of DBs. This proposal also included a WordNet-based NLI to RDBs to access and query DBs using a user's NL [18].

In the same year, **2002**, DBXplorer employed two preprocessing steps, PUBLISH which builds symbol tables and associated structures, and SEARCH which fetches the matching rows from published DBs. Together, PUBLISH and SEARCH enable Keyword-based search to RDBs [12].

In **2004**, PRECISE [47] was developed to use the Semantically Tractable Sentences concept. The semantic interpretation of the sentences in PRECISE is done by analyzing language dictionaries and semantic constraints. PRECISE matches the NLQ tokens with the corresponding DB structures in two stages. Stage 1, it narrows down possible DB matches to the NLQ tokens using the Maximum Flow Algorithm (MFA). MFA finds the best single Flow Network as a directed graph to finally specify one source and one path to increase the flow's strength. MFA returns the maximum # of keywords back to the system. Stage 2 is analyzing the sentence syntactic structure. After that, PRECISE uses the information returned from both stages to accurately transform the NLQ to an equivalent SQL query. However, PRECISE has a poor knowledge base as it only retrieves results that are keyword-based due to NL general complexity and ambiguity.

In **2005**, an NLP question answering system on RDB was defined for NLQ to SQL query analysis and processing to improve the work on XML processing for the structural analysis with DB support. Query mapping was used to derive the DB results [48].

In **2006**, NUITS system was implemented as a search algorithm using DB schema structure and content-level clustering to translate a conditional keyword-based query and retrieve resulting tuples [49].

The Mayo Clinic information extraction system [50] extracts information from free text fields (i.e., clinical notes), including named entities (i.e. diseases, signs, symptoms, procedures, etc.) and their related attributes (i.e. context, status, relatedness to patients, etc.). The system works by implementing simple NLP tasks such as text tokenization, lemmatization, lexicon verification, and POS tagging.

English Wizard, developed in **2009**, is an English NLQ translation tool for querying RDB using SQL. English Wizard distinguishing features include having a graphic user interface (GUI) to issue NLQs or report query results through client/server applications or other DB reporting tools [51].

GINLIDB, in **2009**, employed two types of semantic grammar to support a wide range of NLQs. The first type is a single lexicon semantic grammar for lexicon nonterminal words, and the second type forms terminal phrases or sentences using a composite grammar. In addition, GINLIDB uses ATN for Syntactic analysis to assure tokens structures' compatibility with allowable grammatical structures [52].

In **2010**, CHAT-80 system used Prolog to process NLQs into three stages of representations. First, the system represents NLQ words in their logical constants, and then represents NLQ's verbs, nouns and adjectives with their prepositions as predicates. For NLQ's complex phrases or sentences, the system represents them by forming conjunctions of predicates on the DB. CHAT-80 integrates WordNet as a lexicon and the ontology as the semantic interpreter's knowledge base. CHAT-80 is considered a domain-independent and portable NLIDB. It also uses OWL Ontology to define RDB Entity-Relational model to increase the accuracy of NLQ sentences [53].

In **2012**, an NLI for RDB system called Natural Language Application Program Interface (LANLI) was developed [54]. LANLI utilized semantic and syntactic tree generation for query execution defined on a Local Appropriator. LANLI does the NLQ words matching with the corpus tokens via a matching algorithm before using the query formulating tree. LANLI is an effective NLIDB retrieval system due to its use of accurate tree formation algorithm for both the NLQ and the built-in DB, besides the use of knowledge dictionary-like tables that works at interpreting the knowledge that the NLQs might have.

In **2014**, restricted NLQ to SQL translation algorithm was proposed through CliniDAL [55] focusing on managing the complexity of data extraction from Entity-Attribute-Value (EAV) design model. The algorithm starts with the Controller receiving an NLQ with its main parameters' quantity to perform advanced search. The NLQ gets parsed, categorized and optimized via the Query Processor using the query (parse) tree and the information stored in the data source context model, which is a kind of metadata, to produce a query or parse tree including entities, comparatives and their categories (i.e. Patient, Medical, Temporal, etc.). CliniDAL's grammatical parser recognizes the part of the Restricted NLQ via a free-text resolution mechanism in addition to the data analytics post-processing steps for a question deep analysis. The third step is the Query Translator that identifies the NLQ tokens in the query tree so the Mapper can map them to their internal conceptual representation in the Clinical Information System (CIS). This mapping process is done using the similarity-based Top-k algorithm, in addition to embedded NLP tools (e.g., tokenization, abbreviation expansion and lemmatization). The mapping results are stored in a generic context model which feeds the translation process with necessary information just like an index. As such, the corresponding tables and fields for the NLQ tokens are extracted from the data source context model to be fed into the SQL SELECT clause, and the value tables are extracted from the SELECT clause to generate an SQL FROM clause. The query (parse) tree leaf nodes represent query constraint categories which reflect conjunction or disjunction using algebraic computations for clinical attributes and their values. CliniDAL uses the unique unified term, TERM_ID, and its synonyms as the internal identifier for software processing purposes, as in composing SQL statements.

In **2017**, the special purpose CliniDAL [42] was introduced to integrate a concept-based free-text search to its underlying structure of parsing, mapping, translation and temporal expression recognition, which are independently functioning apart from the CIS, to be able to query both structured and unstructured schema fields (e.g. patient notes) for a thorough knowledge retrieval from the CIS. This translation is done using the pivoting approach by joining fact tables computationally instead of using the DBMS functionalities. Aggregation or statistical functions require further post-processing to be applied and computed.

In **2018**, an NLI to RDB called QUEST [56] was developed on top of IBM Watson UIMA pipeline (McCord) and Cognos. QUEST emphasized focus on nested queries, rather than simple queries, without restricting the user with guided NLQs. QUEST workflow consists of two major components, QUEST Engine Box and Cognos Engine Box. Quest Engine Box includes the schema-independent rule templates that work by extracting lexicalized rules from the schema annotation file in the online part of this box, besides the rule-based semantic parsing module that generates lexicalized rules used in the semantic parsing. Quest Engine Box also includes a semantic parser built on many Watson NLP components, including English Slot Grammar (ESG) parser, Predicate Argument Structure (PAS), and Subtree Pattern Matching Framework (SPMF). This box will finally produce a list of SQL sub-queries that are later fed into the QUEST other box, the Cognos Engine Box. The latter box focuses on final SQL statement generation and execution on IBM DB2 server. QUEST proved to be quite accurate compared to previous similar attempts.

Despite the success of above attempts on NLIDB, token-based [57], form/template-based [58], menu-based [59] or controlled NL-based search, are similar but simpler approaches require much more effort as the accuracy of the translation process depends on the accuracy of the mapping process.

ML Algorithms

NLP tools and ML techniques are among the most advanced practices for information extraction at present [42]. In respect of ML techniques, they could be either *rule-based* or *hybrid approaches* for features identification and selection or rules classification processes. A novel supervised learning model was proposed for i2b2 [60], that integrates rule-based engines and two ML algorithms for medication information extraction (i.e. drug names, dosage, mode, frequency, duration). This integration proved efficiency during the information extraction of the drug administration reason from unstructured clinical records with an F-score of 0.856.

ML algorithms have a wide range of applications in dimensionality reduction, clustering, classification, and multilinear subspace learning [61, 62]. In an NLP, ML is used to extract query patterns to improve the response time by creating links between the NLP input sources and prefetching predicted sets of SQL templates into a temporary cache memory [63]. ML algorithms are typically bundled in a library and integrated with query and analytics systems. The main ML properties include scalability, distributed execution and lightweight algorithms [62].

NLP and knowledge-based ML algorithms are used in knowledge processing and retrieval since the 1980s [64]. Boyan et al. [65] optimized web search engines by indexing plain text documents, and then used reinforcement learning techniques to adjust documents' rankings by propagating rewards through a graph. Chen et al. [66] used inductive learning techniques, including symbolic ID3 learning, genetic algorithms, and simulated annealing to enhance information processing and retrieval and knowledge representation. Similarly, Hazlehurst et al. [67] used ML in his query engine system to facilitate automatic information retrieval based on query similarity measures through the development of an Intelligent Query Engine (IQE) system.

Unsupervised learning by probabilistic Latent Semantic Analysis is used in information retrieval, NLP, and ML [68]. Hofmann used text and linguistic datasets to develop an automated document indexing technique using a temperature-controlled version of the Expectation Maximization algorithm for model fitting [68]. Further, Popov et al. introduced the Knowledge and Information Management framework for automatic annotation, indexing, extraction and retrieval of documents from RDF repositories based on semantic queries [69].

Rukshan et al. [70] developed a rule-based NL Web Interface for DB (NLWIDB). They built their rules by teaching the system how to recognise rules that represent several different tables and attributes in NLWIDB system, what are the escape words and ignore them, in addition to DB data dictionaries, Rules for the aggregate function MAX, and rules indicating several different ways to represent an 'and' or 'as well as' concept, or interval 'equal' concept. Data dictionaries are often used to define the relationship between the attributes to know, for example, which attribute comes first in a comparative operation, and which falls afterwards in the comparative structure. This NLWIDB is similar to the current research idea, however, we intend on building a simpler and more generic algorithm that could be applied on various domains systems. Our algorithm does not use DB elements as the basis or rules identification, as in Rukshan et al.'s research; rather, it uses general sentence structure pattern recognition to form an equivalent SQL statement.

SPARK [71] maps query keywords to ontology resources. The translation result is a ranked list of queries in an RDF-based Query Language (QL) format, called SPARQL, created in SPARK using a probabilistic ranking model. The ontology resources used by SPARK are mapped items represented in a graph format used to feed SPARQL queries.

Similar to SPARK, PANTO [60] translates the keyword-based queries to SPARQL, but PANTO can handle complex query keywords (i.e. negation, comparative and superlatives). Also, PANTO uses a parse tree, instead of graph representation, to represent the intermediate results to generate a SPARQL query.

i2b2 medication extraction challenge [60] proposed a high accuracy information extraction of medication concepts from clinical notes using Named Entity Recognition approaches with pure ML methods or hybrid approaches of ML and rule-based systems for concept identification and relation extraction.

Keyword++ framework [43] improves NLIDB and addresses NLQs' incompleteness and imprecision when searching a DB. Keyword++ works by translating keyword-based queries into SQL via mapping the keywords to their predicates. The scoring process is done using deferential query pairs.

Keymantic system [72] handles keyword-based queries over RDBs using schema data types and other intentional knowledge means in addition to web-based lexical resources or ontologies. Keymantic generates mapping configurations of keywords to their consequent DB terms to determine the best configuration to be used in the SQL generation

HeidelTime [73, 74] is a temporal tagger that uses a hybrid rule-based and ML approach for extracting and classifying temporal expressions on clinical textual reports which also successfully solved the i2b2 NLP challenge.

CliniDAL [27] composes Restricted Natural Language Query (RNLQs) to extract knowledge from CISs for analytics purposes. CliniDAL's RNLQ to SQL mapping and translation algorithms are enhanced by adopting a temporal analyzer component that employs a two-layer rule-based method to interpret the temporal expressions of the query, whether they are absolute times or relative times/events. The Temporal Analyzer automatically finds and maps those expressions to their corresponding temporal entities of the underlying data elements of the CIS's different data design models.

Similar to CliniDAL, TimeText [75] is a temporal system architecture for extracting, representing and reasoning temporal information in clinical textual reports (i.e., discharge summaries). However, it only answers very simple NLQs.

NLQ to SQL Mapping

Giordani and Moschitti [76] designed an NLQ translation system that generates SQLs based on grammatical relations and matching metadata using NL linguistic and syntactic dependencies to build potential SELECT and WHERE clauses, by producing basic expressions and combining them with the conjunction or negation expressions, and metadata to build FROM clauses that contain all DB tables that S and W clauses refer to, supported by pairing with highest-weight meaningful joins, with MySQL framework in the back end. However, queries that involve less joins and SQLs embedding the most meaningful referenced tables are preferred. Those clauses are then combined using a smart algorithm to form the final list of possible SQL statements that have matching structure and clauses' components related to the DB metadata, NLQ tokens and their grammar dependencies, which are mapped with NLQ tokens using a mapping algorithm. Generated SQLs have a weighting scheme which relies on how many results are found, to order them based on probability of correctness. NLQ tokens' textual relationships are represented by the typed dependency relations called the Stanford Dependencies Collapsed (SDC). SDC works by representing the NLQq by its typed SDCq list, which is prepared by first pruning out the NLQ relations of useless stop/escape words, then stemming/lemmatizing the remaining NLQ's grammatical relations to reach the optimized list SDCopt used to build the clauses SELECT S and WHERE W, and lastly adding the relations' synonyms to the SDCopt list. After that, an iterative algorithm q is applied, which adds the modified stems to Π and/or Σ (e.g. subject or object) categories to search the DB metadata for matching fields with weighted projection-oriented stems and generate the SQL clauses S, F, W, or the nested queries, so the answers set A = SELECT S × FROM F × WHERE W contains all potential SQLs related to q. At the end, the system will select the single SQL from the A set that maximizes the probability of answering query Q correctly. This NLIDB system effectiveness and accuracy at selecting the correct SQL depends on the order of the SQL in the generated list. SQLs on top of the list (top 10) have 81% correct data retrieval, and 92% on the top 5 SQLs. Nevertheless, authors believe that these accuracies can be improved by learning a reranker to reorder the top 10 SQLs. Yet, this NLIDB is considered novel as it is expert-independent since all needed knowledge is already in the DB metadata stores [76].

A similar NLIDB system was designed by [77] with comparable performance despite using different approaches as an expert user who specified semantic grammars is used to enrich the DB metadata and also implemented ad hoc rules in a semantic parser [77]. Other similar work is the KRISP [78] system, which achieved 94% Precision and 78% Recall of correctly retrieved SQL answers.

Giordani and Moschitti [24] have innovated a novel model design for automatic mapping of NL semantics into SQL-like languages by doing the mapping at syntactic level between the two languages. After that, Support Vector Machines (SVM) ML algorithm is applied on the mapping results to derive the common high-level semantics to automatically translate NLQs into SQLs. To do this, syntactic parsers were used to define NLQ and SQL trees through the ML algorithm using a statistical and shallow model. SVM is used here to build the training and test sets, where the ML model input is a corpus of questions and their correct answers. SVM then automatically generates the annotated and labeled set of all probable correct and incorrect relational Question/Answer (Q/A) pairs. Those RDB pairs are encoded in SVMs using Kernel Functions to compute the number of common substructures between two trees and produce the union of the shallow feature spaces of NLQs and SQLs. Moreover, Kernel Functions is also a combination of Tree Kernels (e.g. Syntactic Tree Kernel (STK) and Its Extension with leaf features (STKe)) applied to syntactic trees and Linear Kernels applied to bag-of-words, and both applied to the syntactic trees of NLQs and SQLs to train the classifier over those pairs to select the correct SQLs for an NLQ. Then, map this new NLQ to the set of available SQLs and rank all available SQLs according to their classifier scores and only use the higher scores NLQs. Ranking potential SQLs to a given NLQ is done through an SVM using advanced kernels to generate a set of probable NLQ/SQL pairs and classify them to correct or incorrect using an automatic categorizer on their syntactic trees by applying Charniak's syntactic parser on NLQs and a modification of the SQL derivation tree using an ad hoc parser on SQL queries. Then, the top-ranked pairs are selected according to the automatic categorizer probability score output. Authors tested the mapping algorithm of NLQs into SQLs using a standard 10-fold cross-validation, the standard deviation, the learning curve and the average accuracy of correct SQLs selection for each NLQ. This approach proved to be able to capture the shared semantics between NLQs and SQLs. It also proved that the implemented kernel improves the baseline model (32%) according to the cross-validation experiments by choosing correct SQLs to a certain NLQ. However, a polynomial kernel (POLY) of 3rd degree on a bag of words is better than STK because it consists of individual tokens that does not exist in STK. Overall, kernel methods are reliable in describing relational problems by means of simple building blocks [24].

Tseng and Chen [79] aim at validating the conceptual data modeling power in the NLIDB area via extending the Unified Modeling Language (UML) [80, 81] concepts using the extended UML class diagram's representations to capture and transform NLQs with fuzzy semantics into the logical form of SQLs for DB access with the help of a Structured Object Model (SOM) representation [82] that is applied to transform class diagrams into SQLs for query execution [50]. This approach maps semantic roles to a class diagram schema [80, 81, 83] and their application concepts, which is one of the UML 9 diagrams used to demonstrate the relationships (e.g. Generalization and Association) among a group of classes. Carlson described several constraints to build semantic roles in English sentences [84].

UML is a standard graphical notation of Object-Oriented (OO) modeling and information systems design tool used for requirement analysis and software design. UML class diagrams are used to model the DB's static relationships and static data models (the DB schema) by referring to the DB's conceptual schema. SOM methodology is a conceptual data-model-driven programming tool used to navigate, analyze, and design DB applications and process DB queries [79].

Authors of [79] aim to explore NLQ constructs' relationships with the OO world for the purpose of mapping NLQ constructs that contain vague terms specified in fuzzy modifiers (i.e. 'good' or 'bad') into the corresponding class diagrams through an NLI, to eventually form an SQL statement which, upon execution, delivers answers and a corresponding degree of vagueness. Authors focused on the fuzzy set theory [49] because it is a method of representing vague data with imprecise terms or linguistic variables [85, 86]. Linguistic variables consist of NL words or sentences (i.e. old, young), excluding numbers (i.e. 20 or 30), yet imprecise NLQ terms and concepts can be precisely modeled using these linguistic variables by specifying natural and simple specifications and characterizations of imprecise concepts and values.

In [79] real-world objects' connectivity paths are mapped to SQLs during the NLQ execution by extracting the class diagram from the NLQ in a form of a sub-graph/tree (a validation sub-tree) of the SOM diagram that contains relevant objects connecting the source and the target, that have been identified by the user earlier in a form of objects and attributes. The source is objects and their associations that have valued attributes to illustrate the relationship of objects and attributes of interest, while the object of ultimate destination is the target. Results are then sent to the connectivity matrix to look for any existing logical path between the source and the target to eventually map the logical path to an equivalent QL statement which can be simplified by inner joins. Schema and membership function represented in class diagram are used to link each fuzzy modifier with their corresponding fuzzy classes.

Isoda [87] discovered that OO-based analysis applications enable intuitive and natural real-world modelling by identifying the corresponding classes of those real-world objects.

Moreno and Van [88] proved that NLQs constructs' conceptual modeling formalization can be mapped naturally into an OO conceptual model, just like how Metais [89] mentioned about NLQ and DB conceptual schema, and how efficient they are at representing the real world's conceptualization features.

Yager et al. [90] prove that the fuzzy sets (consists of fuzzy terms like young and rarely) theory provides a linguistic-based application of NLQs modeling. Furthermore, fuzzy NLQs allow users to describe real-world objects more intuitively through vague predicates including larger number of tuples.

Based on [79], fuzzy NLQs with linguistic terms and fuzzy terms are more flexible compared with precise NLQs as they provide more potential answers in case of no available direct answers. Regarding the linguistic inter-relationship with DB schema, NLQs are linguistically analyzed to reduce ambiguity and complexity by using a linked dictionary and predefined grammar-based rules, while a DB schema acts as a DB conceptual design blueprint.

In the same vein, Owei [91] came up with a concept-based QL that facilitates query formulation by means of DB queries' conceptual abstraction to exploit the semantic data models and map NLQ's constructs to their equivalent specific objects in the real-world DB.

L2S [92] is a hybrid approach to transform NLQs into SQL. It maps NL vocabularies to SQL using semantic information via underlying tools and uses bipartite tree-like graph-based processing models for handling and evaluating the remaining lexicons for the matching stage. This hybrid approach was designed to help language transformation systems that lack adequate training data and corpus, specific domain background knowledge and observation analysis and inefficient employed NLQ linguistic tools (e.g., tokenizer and parser).

Most language transformation approaches or question answering systems rely on rich annotated training data with employed statistical models or non-statistical rule-based approaches. Generally speaking, rule/grammar-based approaches [93] require extensive manual rules defining and customizing in case of any DB change to maintain accuracy and is used more often in real-life industrial systems [94].

Statistical models are mostly used in academic research, and they work by building a training set of correct and incorrect NLQ to SQL pairs [24]. The language transformation then becomes a binary classification task to correctly map and label each NLQ to its equivalent SQL. In addition, ML Features extraction is derived from tokens and syntactic trees of correctly mapped and labeled NLQs into SQLs transformation processes [95].

In addition to statistical and non-statistical models, there are syntactic-based analysis [59] that is not rule-based, as well as graph-based models [47] and Prolog-like representation [96], in addition to complex NLQ to SQL transforming operations in frameworks such as the high-level ontology data representation with huge amount of data.

L2S functions are spread across three steps [92]. During the first preprocessing step, NLQ and DB are analyzed, and all ML features are extracted. This step is supported by three pre-processing components: a linguistic component that analyzes the NLQ to generate the list of lexicons and discover any conditional tokens (i.e. greater/less than). This component has embodied tools such as Name Entity recognizer (NER) [97] and Coltech-parser in GATE [98] to extract semantic information that lies within an NLQ. Also, this component handles the NLQ tokenization and named entity tagging by using Java Annotation Patterns Engine (JAPE) grammars. The output of this first step is an attachment constraint. The second component is the lexicon, used for DB elements analysis and their attachment constraints. The third component is the ambiguity Solving, used for NLQ input correction to guarantee an ambiguity-free NLQ. In this Component, L2S compares all NLQ tokens with DB elements to find out the ambiguous tokens that match with non or more than two DB elements through the use of ellipsis method or the highest possibility selection. The next L2S step is the Matching step which has two main components, the Semantic Matching and the Graph-Based Matching components. This step handles the interlingual mapping to produce equivalent SQL elements to the output of the first step. The third step is generating a complete and accurate SQL statement. Authors of [92] built a mapping table manually to match the DB lexicons with the NLQ tokens. L2S transforms NLQ and DB elements into a tree-like graph, then extracts the SQL from the maximum bipartite matching algorithm result. L2S effectiveness was validated and the results maintained high accuracy over different domains. It was also proved that switching domains requires minimal customization work.

According to [99], NLP is today an active technique of Human–Computer Interaction, especially in the social media era [99], developed a structural design method to automatically convert and translate simple DDL and DML queries with standard join conditions from an NLQ format into SQLs through NLQ's semantic extraction and optimized SQL generation. This work also provides a user-friendly NLI for end users to easily access the social web DB via any web source. Authors used Java Programming Language and its technical tools to build the NLIBD system's front-end and used R-tool as a data collector to gather data from social web sources. For data storage in the system's backend, authors used an oracle DB, the SQL server. They also used a limited Data Dictionary to store all system-related words. The system would receive an NLQ, process it, collect data using the R-tool interface, extract the semantic knowledge from the social web source, and finally generate the associated SQL statement.

In [99], a simple architectural layout framework is proposed. It starts with the four pre-processing phase modules; Morphological Analysis Module, Semantic Analysis Module, Mapping Table Module, and the Reports' Retrieval Module. In the first module, the Morphological Analysis, NLQ is received and tokenized into words, which will be then passed on to the extractor to stem them by identifying their root words. At this stage, unwanted words are removed. Then, the extractor does the stemming processes using the Porter algorithm, in addition to maintaining previously tokenized words, a.k.a. the predefined words, from previous NLQs to compare them with the newly tokenized words to extract the main keywords, which are passed to the second module. NLQ tokens' synonyms are identified from the integrated DB elements' names (i.e., column or table names), which are used to replace the extracted keywords.

The Semantic Analysis Module generates a parse tree from the identified keywords and passes it on to the third module. The Mapping Table Module has all potential SQL templates and knows the maximum possibility of each NLP word, and hence does all the identified words mapping using the mapping table. The best suitable query is generated and passed on to the fourth and last module, the Reports' Retrieval Module, to deliver it to the end user as a report.

The noticeable effort in this work is that every entered NLQ goes through a Syntactic Rules Checker for any grammatical mistakes. Also, semantic analysis is used to map NLQ tokens to DB objects. The combination of tokens' meanings defines the NLQ general meaning, which is used to come up with a list of potential SQL queries among which the end user has to choose.

Akshay [100] proposes an NLP search system interface for online Applications, search engines and DBs requiring high accuracy and efficiency.

Kaur [101] illustrates the useful usage of Regular Expressions (regexps), which are generic representations for strings, in NLP phonology and morphology, text search and analysis, speech recognition and information extraction. However, clear collections of regexps in NLQ sentences are not clearly specified.

Avinash [102] uses domain ontology in NLIDBs for NLQ's semantic analysis and emphasizes on employing domain and language knowledge at the semantic level to enhance precision and accuracy.

Kaur and Bali [45] examined an NLQ into SQL conversion interface module by means of NLQ's syntactic and semantic analysis, but this module is unable of processing complete semantic conversion for complex NLQ sentences.

Arati [103] used Probabilistic Context Free Grammar (PCFG) as an NLDBI system design method for NLP. It works by using NLQ's syntactic and semantic knowledge to convert NLQ into an internal representation and then into SQL via a representation converter. However, finding the right grammar for optimization is challenging in this system design.

Dshish [104] used NLP for query optimization to translate NLQ into SQL by means of semantic grammar analysis, while using the LIFER/LADDER method in the syntax analysis. Since the LIFER/LADDER system only supports simple SQLs formation, this translation architecture is largely restricted.

For RDBMSs, Gage [105] proposed a method of an AI application in addition to fuzzy logic applications, phrase recognition and substitution, multilingual solutions and SQL keyword mapping to transform NLQs into a SQLs.

Alessandra [24] used Syntactic Pairing for semantic mapping between NLQs and SQLs for eventual NLQ translation using SVM algorithm to design an RDB of syntax trees for NLQs and SQLs pairs, and kernel functions to encode those pairs.

Gauri [106] also used semantic grammar for NLQ into SQL translation. In the semantic analysis, the author used the Lexicon to store all grammatical words, and the post-preprocessor to transform NLQs' semantic representations into SQL. However, this architecture can only translate simple NLQs, but not flexible.

Karande and Patil [51] used grammar and parsing in an NLIDB system for data selection and extraction by performing simple SQLs (i.e. SQL with a join operation or few constraints) on a DB. This architecture used an ATN parser to generate parse trees.

Ott [107] explained the process of SQLs Automatic Generation via an NLIDBs using an internal intermediate semantic representation language based on formal logic of the NLQs that is then mapped to SQL+. This approach is based on First Order Predicate Calculus Logic resembled by DB-Oriented Logical Form (DBLF), with some SQL operators and functions (e.g., negation, aggregation, range, and set operator for SQL SELECT).

This approach, called SQL+, aims to solve some of the SQL restrictions such as handling ordinals via loop operators (e.g., the 6th lowest, the 3rd highest). To replace the loop operator, SQL + expressions are entered into a programming interface to SQL supplied with a cursor management.

SQL+ strives to save the ultimate power of NLQ by augmenting the SQLs in a way that each NLQ token is represented and answered by SQL expressions. Experiment results prove that even complex queries can be generated following three strategies, the Join, the Temporary Relation and the Negation Strategy, in addition to a mixture of these strategies [107].

For the join strategy, and in the DBLF formula, for each new relation reference a join operation is built in the SQL FROM clause recursively in a top-down direction. Universal quantifiers are usually implemented by creating counters using double-nested constructs as in (NOT EXISTS [sub-SQL]) which has been used in the TQA system [108]. However, [107] uses the temporary relation creation strategy instead to handle universal and numeric quantifiers and to handle ordinals and a mixture of aggregate functions as well. The temporary relations are created using the join-strategy to easily embed them in any SQL expression. Hence, whenever there is a quantifier, a temporary relation is built for it recursively. For the negation strategy, and in DBLF, negation is handled by setting the "reverse" marker for yes/no questions if the negation at the beginning of the sentence, and by using (NOT IN [subquery]) constructs in case of verb-negation and negated quantifiers in other positions. Both negation handling methods are doable if the negation occurs in front of a simple predicate, and in this case, the number and position of negation particles is not restricted. For the Mixed strategies, any of the previous three strategies can be mixed arbitrarily as in building temporary relations when aggregate functions or ordinals occur.

TQA [108] is an NLI that transforms NLQs into SQL directly using semantic grammar and deep structure grammar to obtain a higher performance and better ellipses and anaphora handling. However, TQA and similar systems are almost not transportable and barely adaptable to other DB domains.

PHLIQA1 [4], ASK [46] and TEAM [16] adopt the intermediate semantic representation languages connected with a Conceptual Schema (CS) to provide an efficient NLQ into SQL transformation tool. CS helps mapping NLQ tokens to their DB lexicon representations because it stores all NLQ tokens, DB terms of relations and attributes and their taxonomies, in addition to the DB hierarchy structure and metadata.

The USL system [109] is adaptable and transportable because it has a customization device. Yet its intermediate semantic and structure language is syntax-oriented, and not based on predicate logic. Hence, some semantic meanings are represented though tree structures forms.

TQA and USL techniques together form the LanguageAccess system [107], which has a unique SQL generation component that uses the DBLF and the Conceptual Logical Form (CLF) as two distinct intermediate semantic representation languages.

LanguageAccess system works through many steps. First, a phrase structure grammar parses an NLQ to generate parse trees, which are then mapped to CLF using the CS. Generated CLF formulae are paraphrased in NL and then presented to the end user for ambiguous tokens interpretations and meaning verification. Once the end user chooses a CLF formula, using the CS, it gets transformed to DBLF, the source of SQL generation, which is then transformed to SQLs. DBLF considers the DB values internal representations (e.g. strings, numbers), the temporary relations order, and the generated expressions delivery mechanism to DBs.

Authors of [7, 110, 111] used NLQs semantic parsing to model algorithms to map NLQs to SQLs. Similar research work is done by [112] using specific semantic grammar. Authors of [7, 110, 113] used lambda calculus and applied it on NLQs meaning representation for the NLQ to SQL mapping process. Furthermore, [114] used ILP framework's defined rules and constrains to map NLQs using their semantic parsing. For the same purpose, [7, 110, 111] followed a time-consuming and expensive approach by producing NLQ tokens' meaning representations manually. Similarly, [112] developed an authoring system through extensive expertise time and efforts on semantic grammar specification. Authors of [7, 110, 113] developed a supervision-extensive system using lambda-calculus to map NLQs to their corresponding meaning representations.

Similar to KRISP [78], Giordani and Moschitti [115] developed a model using only Q/A pairs of syntactic trees as the SQL compiler provides the NLQs derivation tree that is required to translate factoid NLQs into structural RDB SQLs with generative parsers that are discriminatively reranked using an advanced ML SVM-ranker based on string tree kernels. The reranker reorders the potential NLQ/SQL pairs list which has a recall of 94%, recalling the correct answers in this system.

The system in [115] does not depend on NLQ-annotated meaning resources (e.g. Prolog data, Lambda calculus, MR, or SQLs) or any manual semantic representations except for some synonym relations that are missing in WordNet. The first phase is the generation phase where NLQ tokens' lexical dependencies and DB metadata-induced lexicon in addition to WordNet are used, instead of a full NLQ's semantic interpretation, to build the SQL clauses (i.e. SELECT, WHERE, FROM, joins, etc.) recursively with the help of some rules and a heuristic weighting scheme. DB metadata does the relations disambiguation tasks and includes DB data types, Primary Keys (PKs) and Foreign Keys (FKs) and other constraints, names of entities, columns and tables according to domain semantics, and is also called DB catalog usually stored as INFO_SCHEMA (IS) in a DB. The output of the generation phase is a ranked potential SQLs list created by the generative parser.

In Dependency Syntactic Parsing is used to extract NLQ tokens' lexical relations and dependencies. According to [19], WordNet is efficient at expanding predicate arguments to their meaning interpretations and synonyms; however, WordNet generalizes the relation arguments but does not guarantee NLQ's lack of ambiguity and noise which affects its meaning interpretation significantly. Therefore, this system generates every possible SQL with all of its clauses, including ambiguous ones, based on NLQs lexical and grammatical relations dependencies matches, extracted by the Stanford Dependencies Parser [116], and SQL clauses' logical and syntactic formulation structures.

The first relation executed on the GEOQUERIES corpus in the [115] algorithm is the FROM clause relation to find the corresponding DB tuples considering the optional condition in the WHERE clause and then match the results with the SELECT clause attributes. In case of any empty clauses or nested queries mismatching, this algorithm will generate no results; otherwise, correct SQLs are generated among the top three SQLs in 93% of the times using standard 10-fold cross-validation performance measure. This high accuracy and recall are due to the robust and heuristic weights-based reranker that is built using SVM-Light-TK6 extending the SVM-Light optimizer [117] by employing the tree kernels [118, 119] to use the addition STKn + STKs or the multiplication STKn × STKs. Default reranker parameters are used such as in the normalized kernels, $\lambda = 0.4$ and cost and trade-off parameters = 1. However, this approach mandates the existence of possible SQLs in advance as no new SQLs can be generated by the algorithm, it only verifies if an entered NLQ has a corresponding SQL to produce a correct answer.

Conceptually similar to [115], Lu et al.'s [120] mapping system does not depend on NLQ annotation either, but on a generative model and the (MODELIII+R) which is a discriminative reranking technique. Also, DCS system [121] does not depend on a DB annotation either and works as well as a mapping system enriched with prototype triggers (DCS+). In addition, from Q/A pairs, SEMRESP employs a semantic parser learner [122] that works best on annotated logical forms (SQLs). Kwiatkowski et al. [123] developed UBL system that when trained on SQLs and Q/A pairs, it is able to use restricted lexical items together with some CCG combinatory rules to learn newly entered NLQ lexicons.

CURRENT RESEARCH WORK JUSTIFICATION

The main goal proposed in this work is to find a simple but accurate mapping mechanism between NLQ and machine-readable instructions such as the RDB query language, SQL. To date, there is no adequate NLQ into SQL translation mechanism that does not compromise accuracy and precision with complexity or exaggerated simplicity to an unfunctional level. Such translation mechanisms have numerous rules exceptions and resulted errors when applied on other RDBs. Thus, the proposed research exploits a simple manually written rule-based mapping constraints algorithm as a design to a unique and accurate NLQ into SQL translation mechanism. This algorithm maps NLQ tokens semantic and syntactic information into RDB elements categories (i.e., value, attribute, etc.) and then into SQL clauses consistently. The algorithm uses computational linguistics analysis pairing and matching mechanisms through MetaTables. The study of translating NLQ into SQL-like languages has a long history starting from the 1971 to date [2–4, 12, 16–18, 35, 39, 40, 41, 46–49, 52, 53, 57, 59, 60, 65–69, 70–73, 99, 100, 124–130].

According to the literature [57, 74], mapping NLQ into SQL occurs using any of the following approaches:

Authoring Interface-Based Systems [131]

By using semantic grammar specifications designed by extensive expertise time and efforts to identify and modify RDB elements and concepts (e.g., CatchPhrase Authoring tool [131]).

Enriching the NLQ/SQL Pair

By adding extra metadata to the pairs to easily find a semantic interpretation for NLQ's ambiguous phrases for the matching problem (e.g., Inductive Logic Programming (ILP) [75]).

Using MLA Algorithms

By using correct NLQ/SQL pairs' corpora. A corpus induces semantic grammar parsing to map NLQs into SQLs by training a Support Vector Machine (SVM) classifier [33] based on string subsequence kernels (i.e., Krisp [132]).

Restricted NLQ Input [35]

By using a keyword-based search structure [128] through a form, template or a menu-based Natural Language Interface (NLI) [52] to facilitate the mapping process.

Lambda Calculus [7, 112, 114]

Applied on NLQs meaning representation for the NLQ into SQL mapping process.

Tree Kernels Models [29, 47, 78, 100, 126]

A Kernel Function [110, 113, 133] is a combination of Tree Kernels [134] such as the Polynomial Kernel (POLY) [135], Syntactic Tree Kernel (STK) [33] and its extension with leaf features (STKe) [132]. They can be applied on NLQs/SQLs pairs syntactic trees, while Linear Kernels [136] are applied on a "bag-of-words". They are used to train the classifier over those pairs to select a correct SQL for a given NLQ.

Unified Modeling Language (UML) [34, 118]

A standard graphical notation of Object Oriented (OO) Modeling [137] and an information system design tool. UML is a combination of Rumbaugh's Object-Modeling Technique (OMT) [119], Booch's OO Analysis and Design [74], and Jacobson's Objectory [100]. UML class diagrams are used to model the DB's static relationships and static data model (DB schema) by referring to the DB's conceptual schema.

Weighted Links [81]

Is a mapping system that works by pairing with the highest weight meaningful joins between RDB lexica and SQL clauses.

What follows is an explanation of each of the above mapping approaches and a justification of excluding them by choosing the second approach, enriching the NLQ/SQL pairs, as the most effective one.

In the current research, the second approach is adopted because it does not restrict the user to using certain domain-specific keywords [40, 72], as is the case in the restricted NLQ input approach. This is because the aim in the current work is to facilitate the HCI without users' prior knowledge of the RDB schema and the system's infrastructure, underlying NLP linguistic tools or any specific query language.

In addition, NLQ is the most natural way of communication for humans. The first approach, authoring interface-based systems, relies heavily on end-user input through all the interface screens that they have to go through to specify and modify the used keywords or phrases. Hence, it might seem to the user that it would have been easier for them if they had any programming knowledge to enter the SQL statement directly without using the Natural Language Interface (NLI) screens. Hence, the current work only involves end users in the case of any spelling mistakes or ambiguate phrases. In this regard, in [40, 52, 68, 69, 87], authors described the menu-based or restricted keyword-based NLQ approaches as methods of mapping. In their paper, they explained how insignificant restricted NLQ input systems are in terms of accuracy and recall. Besides, it also has portability problems even with advanced algorithms such as Similarity-Based Top-k Algorithm [79] that compares the similarity between dictionary k-records and NLQ tokens. This algorithm achieved an accuracy of 84% only [100], whereas the current research system achieved an accuracy of as high as 95%.

Most language translation approaches or QAS systems rely on rich annotated training data (corpus) with employed statistical models or non-statistical rule-based approaches. As such, the third approach that relies on MLA algorithms requires the presence of huge domain-specific (specific keywords used in the NLQ) NLQ/SQL translation pairs' corpora. Such a corpus is difficult to create because it is very time-consuming and a tedious task required by a domain expert. NLQ/SQL pairs corpus requires hundreds of manually written pairs written and examined by a domain expert to train and test the system [70, 76, 138].

Avinash [102] employed a domain-specific ontology for the NLQ's semantic analysis. As a result, Avinash's algorithm would fall under the over-customization problem, making the system unfunctional on any other domain. It is also neither transportable nor adaptable to other DB environments, except with extensive re-customisation. Such domain-specific systems assume the user is familiar with the DB schema, data and contents. On the other hand, the current research work uses simple algorithmic rules and is domain-independent. Hence, it does not assume prior knowledge of the adopted RDB schema or require any annotated corpora for training the system. Instead, it uses linguistic tools to understand and translate the input NLQ. However, the used NLQ/SQL pairs are only used for algorithm testing and validation purposes. Furthermore, relying heavily on MLAs proved to be not effective in decreasing the translation error rates or increasing accuracy [139]. This remains the case even after supplying the MLA algorithm with a dedicated Error Handling Module [77]. In this regard, the current research work took proactive measures by using NLP linguistic techniques to make sure the NLQ is fully understood and well interpreted. This full interpretation happens through the intermediate linguistic layers and the RDB MetaTable before going any further with the processing; to avoid potential future errors or jeopardize accuracy. Computational linguistics is used here in the form of linguistics-based mapping constraints using manually written rule-based algorithms. Those manually written algorithms are mainly observational assumptions summarised in Table 4 (Chapter 4). Table 4 specifies RDB schema categories and semantic roles to map the identified RDB lexica into the SQL clauses and keywords.

Generally speaking, rule/grammar-based approaches [102] require extensive manual rules defining and customizing in case of any DB change to maintain accuracy [140]. However, the rule-based observational algorithm implemented in the current research work is totally domain-independent and transportable to any NLQ translation framework. Generally, mapping is a complicated science [14] because low mapping accuracy systems are immediately abandoned by end users due to the lack of system reliability and trust. Hence, this research work proposes a cutting-edge translation mechanism using computational linguistics. However, there are several aspects of the proposed research contribution which will be discussed in reference to the two mapping algorithms in Figure 8 (Chapter 4).

MAPPING NLQ TOKENS INTO RDB LEXICA

NLQ Tokens Extraction

In the current research work, NLQ tokens extraction and their types identification happen through deep computational linguistics processes. The processes are done via underlying NLP linguistic tools that use an English word semantics dictionary (WordNet), RDB MetaTable (for the mapping algorithm) and a mapping table for unique values namely, Primary Keys (PKs) and Foreign Keys (FKs). The adopted linguistic method proved to be more accurate and effective than other tokens extraction methods such as:

- Morphological and word group analyzers for tokens extraction [35],
- Pattern Matching [141] to identify keywords and their types,
- NER Recognizer [93] alone with the Coltech-parser in GATE [94] to tokenize and extract NLQ's semantic information,
- Java Annotation Patterns Engine (JAPE) grammars [88] for NLQ tokenization and named entity tagging,
- Porter algorithm [97] to extract tokens' stems,
- Unification-Based Learning (UBL) algorithm [98] which uses restricted lexical items and Combinatory Categorial Grammar (CCG) rules [98] to learn and extract NLQ tokens,
- Dependency Syntactic Parsing [134] to extract tokens and their lexical relations,
- Dependency-Based Compositional Semantics (DCS) [92] system enriched with Prototype Triggers [92], or
- Separate value and table extractor interfaces [29], which is a compromising approach for not supporting the RDB schema elements' MetaTables and synonyms such as in the current proposed system.

RDB Lexica Mapping

Even recent studies in this field [142] failed to score high accuracy for the tokens mapping algorithm or handling complex SQLs, despite using state-of-the-art tools and techniques. An example of recent works in 2018 is Spider [142], which does its mapping using a huge human labeled NLQ/SQL pairs corpus as a training and testing dataset. Such datasets are created using complex and cross-domain semantic parsing and SQL patterns coverage. However, Spider's performance surprisingly resulted in a very low matching and mapping accuracy. Hence, the current research work is distinct from most of the previous language translation mechanism efforts because the focus here gives highest priority to simplicity and accuracy of the algorithm's matching outcome.

The current research work employs the NLQ MetaTable (Table 1) to map NLQ tokens into RDB lexica. The NLQ MetaTable covers NLQ words, their linguistic or syntactic roles (noun, verb, etc.), matching RDB category (table, value, etc.), generic data type (words, digits, mixed, etc.), unique as PK or FK, besides their synonyms and enclosing source (i.e., tables or attributes). MetaTables are used to check for tokens' existence as a first goal, then mapping them to their logical role as a relationship, table, attribute or value. The general-purpose English language ontology (WordNet) are used to support the MetaTables with words' synonyms, semantic meanings and lexical analysis.

The implemented MetaTables fill up the low accuracy gap in language translation algorithms that do not use any sort of deep DB schema data dictionaries such as [81, 123], or just a limited data dictionary such as [43]. According to [19], WordNet is efficient at expanding NLQ predicate arguments to their meaning interpretations and synonyms. However, WordNet generalizes the relation arguments and does not guarantee NLQ's lack of ambiguity and noise, which significantly affects its meaning interpretation. Hence, supportive techniques are employed in the current research work such as the disambiguation module. In addition, to avoid confusion around the RDB unique values, data profiling [121] is performed on large RDB's statistics to automatically compile the mapping table of unique values, PKs and FKs, based on which RDB elements are queried more often. Mapping tables are manually built for smaller RDBs, while using a data-profiling technique to build them for larger RDBs. Unique values are stored in the mapping table by specifying their hosting sources while a hashing function is used to access them instantly.

RDB Lexica Relationships

NLQ parsing and dependency trees (i.e., Augmented Transition Network (ATN) Lexical Relations Parser [143]) are used in the current research work as part of the NLP semantic and syntactic parsing. Those parsers generate grammar parse trees to explain the NLQ tokens' dependencies and relations [144].

Besides, the RDB lexical join conditions are also discovered between any two words or values. The joint is based on the words' or values' connectivity status with each other or having common parent node in the dependency tree. The parsing helps with the NLQ semantics extraction and RDB lexical data selection. RDB elements relationships are controlled by using only verbs to represent any connectivity in the RDB schema. The verbs' parameters (subject or object) are mapped with the RDB relationship's corresponding elements: tables, attributes or values. If the NLQ verb is unidentified or missing, the relationship between NLQ tokens will be found by analysing the matching RDB lexica intrarelationships with each other.

There are other methods in the literature that identify lexical dependencies and grammatical relations, such as Stanford Dependencies Parser [145], Dependency Syntactic Parser [134] and Dependency-Based Compositional Semantics (DCS) Parser [92]. The current research work used a simple way of representing RDB elements inter-/intra-relationships. This representation restricts the RDB schema relationships to be in the form of a verb for easy mapping between NLQ verbs and RDB relationships.

NLP Syntax and Semantics Definition

The current research discovered common semantics between NLQ and SQL structures by analyzing both languages' syntax roles. In the same vein, understanding the NLQ, by finding the combination of its tokens' meanings, is the most essential part in the language mapping and translation process. Thus, computational linguistic studies at the words processing level is employed as opposed to approaches similar to Lambda Calculus or Tree Kernels Models mentioned above in the fifth and sixth mapping approaches.

The current research framework overcomes any poor underlying linguistic tools' performance that are meant to analyse NLQ syntax and semantics. It overcomes such inadequacies by using the supportive RDB schema knowledge and semantic data models (MetaTables) and WordNet ontology. Furthermore, NLP tools, such as NER tagger, tokenizer or dependency parser, are also employed in addition to the syntactic-based analysis knowledge [52] to generate parse trees from the identified tokens for proper mapping with the related RDB elements. Nevertheless, relying solely on NLQ's syntactic and semantic analysis for the mapping process is not sufficient and produces substantially low precision, False Positive Ratio (FPR) and True Negative Ratio (TNR) as indicated in [1, 7, 46, 51, 57, 76, 115, 116, 131]. Such systems include the LIFER/ LADDER method in [45], SVM algorithm, or SVM-Light optimizer [146], NLQ/SQL syntax trees encoded via Kernel Functions [24] or the Probabilistic

Context Free Grammar (PCFG) method [111] which proved to be challenging in terms of finding the right grammar for optimization. Hence, the current work supports the NLQ's syntactic and semantic grammar analysis with computational linguistics in the form of RDB and NLQ MetaTables.

MAPPING RDB LEXICA INTO SQL CLAUSES

SQL Clauses Mapping

While the current work uses computational linguistics mapping constraints to transform RDB lexica into SQL clauses and keywords, [119] uses the extended UML class diagrams representations [34, 104, 118]. Those representations are used to extract fuzzy tokens' semantic roles, which are imprecise terms or linguistic variables consisting of fuzzy terms like 'young', 'rarely', 'good' or 'bad' [60, 83, 86, 103, 117, 137, 141]. Fuzzy tokens' semantic roles are extracted in the form of a validation sub-graph or tree of the Self Organizing Maps (SOM) diagram representation [147]. SOM diagrams transform UML class diagrams into SQL clauses [148] using the fuzzy set theory [49]. According to [119], fuzzy NLQs with fuzzy linguistic terms are more flexible compared with precise NLQs as they provide more potential answers in the case of no available direct answers. However, though this might provide higher measures of recall, it is significantly compromising the FPR ratio. Hence, fuzzy NLQs are not considered in the current research.

In [119], similar approaches to the current work are implemented for RDB lexica mapping into SQL clauses. The work of [119] uses RDB relationships to map the lexica into NLQs linguistic semantic roles' classes as a conceptual data model. However, since the current work uses supportive NLP linguistic tools, it is more capable of "understanding" the NLQ statement before translating it into SQL query, which highly contributes to the increase in the translation accuracy. Regarding the linguistic inter-relationships within the RDB schema in the current work, not only WordNet is used, but also a Natural Language Toolkit (NLTK) and NLP linguistic tools. In addition, a manual rule-based algorithm is also used to define how NLQ linguistic roles match with the RDB elements, which does not exist in [119] and which explains the variance in translation accuracy and precision in comparison.

Regarding the linguistic analysis used in [119], the user has to identify the fuzzy NLQ source (object), its associations or relationships and the target (attribute) to connect them together for the UML class diagram extraction phase. The results are derived from the connectivity matrix by searching for any existing logical path between the source and the target to eventually map them into an equivalent SQL template. In comparison, and since the current research work aims for a seemingly natural HCI interaction, the user does not have to identify any semantic roles in their NLQ. This is because the underlying NLP tools does this for them. Also, the relationships are identified by the NLQ verbs, so the user is communicating more information in their NLQ using the current research algorithm compared to the other literature works. Hence, it is considered more advanced and user-friendly than that in [119]. Also, not only objects and attributes are extracted from the NLQ; the proposed research work extracts much lower-level linguistic and semantic roles (i.e., gerunds and prepositions) which help select the matching RDB lexica with higher accuracy and precision.

Complexity vs Performance

The current research work is considered significantly simpler than most complex mapping approaches such as [29] as it relies on fewer, but more effective, underlying NLP linguistic tools and mapping rules. An example of a complex language translation model is the Generative Pre-trained Transformer 3 (GPT-3) [30], introduced in May 2020. GPT-3 is an AI deep learning language translation model developed by OpenAI [31]. GPT-3 is an enormous artificial neural networks model with a capacity of 175 billion machine learning parameters [32]. Therefore, the performance and quality of GPT-3 language translation and question-answering models are so high [2]. GPT-3 is used to generate NLP applications, convert NLQ into SQL, produce human-like text and design machine learning models.

However, GPT-3 NLP systems of pre-trained language representation must be trained on text, in-context information and big data (i.e., a DB that contains all internet contents, a huge library of books and all of Wikipedia) to make any predictions [31]. Furthermore, for the model training, GPT-3 uses model parallelism within each matrix multiply to train the incredibly large GPT-3 models [30]. The model training is executed on Microsoft's high-bandwidth clusters of V100 GPUs.

Training on such advanced computational resources would largely contribute to the excellent performance of GPT-3 models. The biggest weakness of this model is its extreme complexity, advanced technology requirements and that it is only efficient once trained because GPT-3 does not have access to the underlying table schema.

Algorithm simplification is necessary in such language-based applications. An example of a complicated system is L2S [92] that compares all existing NLQ tokens with existing DB elements. This approach consumes a lot of time to run through all DB elements to compare them with the NLQ tokens. L2S uses NLP tools, tokens' semantic mapper and graph-based matcher, hence, simplicity is key in the current work. Other examples of complicated systems are in [29] and [24] where a hybrid approach is implemented using Bipartite Tree-Like Graph-Based Processing Model [89], Ellipsis Method [82] and the Highest Possibility Selection Algorithm [50]. Those approaches require a domain-specific background knowledge and a thorough training dataset. Hence, they are not considered in the current research work. The RDB lexica into SQL clauses mapping algorithms in the literature ranged from simple to complex methods.

After thorough study and research, it has become clear that the proposed algorithm in the current research work is the best in terms of performance, simplicity, usability and adaptability to different framework environments and RDB domains. Both implemented mappers have access to an embedded linguistic semantic-role frame schema (WordNet and Stanford CoreNLP Toolkit), MetaTables and the RDB schema and MetaTables. Those resources are essential for SQL templates formation and generation which is a popular problem under the NLP era.

The majority of NLQ into SQL mapping processes employ sophisticated semantic and syntactic analysis procedures on the input NLQ [24, 79, 149–151]. However, those analyses are computationally expensive. Hence, the current research work employs a lightweight approach for this type of query translations. In particular, the use of MetaTables which defines the lexicon semantic role (i.e., noun, verb, etc.) and its adjacent SQL slot, prioritising accuracy above complexity. Complex algorithms, such as the weighted links approach, compromise accuracy for complexity. An example is in [76] that generates ordered and weighted SQLs scheme using Weighted Neural Networks [39] and Stanford Dependencies Collapsed (SDC) [152] as grammatical dependency relations between NLQ tokens. This system is expensive to implement and unscalable to bigger RDBs. It also prioritizes SQLs based on probability of correctness instead of accuracy and precision.

However, what is interesting in [76] is their use of linguistics in their algorithm where they identify NLQ's subject or object to search the DB for matching attributes. This matching uses Weighted Projection-Oriented Stems [127] to generate the SQL clauses accordingly. Yet the translation accuracy of this algorithm still falls behind the proposed algorithm in the current research work. This is because the current work uses further linguistic categories (i.e., adjectives, pronouns etc.) in addition to using the verbs to find the attributes' and values' intra-relationships instead of using a heavy weighted tool such as the Weighted Projection-Oriented Stems in [76].

Photon [153] is another neural network-based NLQ into SQL translation approach that translates NLQs by parsing them into executable SQL queries. Photon is the state-of-the-art NLIDB introduced to the public in June 2020. It employs several modules as sublayers, such as a deep learning-based neural semantic parsing module, an NLQ corrector module, a DB engine and a response generator module. For the neural semantic parsing layer, BERT and a bi-directional Long-Short Term Memories (LSTM) machine learning algorithms were used to produce hidden representations that match NLQ tokens with table and attribute names. The NLQ corrector module detects untranslatable NLQs by highlighting an ambiguous or a confusion span around the token and then asks the user to rephrase the NLQ accordingly. Although Photon represents a state-of-the-art NLIDB and an NLQ translation mechanism, it still falls under the complex translation models while lighter weight translation algorithms are sought for. In addition, the Photon model relies on training datasets to train its translation algorithm, table value augmentation module, static SQL correctness checking module and neural translatability detector module. Furthermore, deep learning and neural networks approaches, generally speaking, tend to act as a black box where it is hard to interpret their predictions and hard to analyse their performance and evaluation metrics.

An example of a simple mapping algorithm is SAVVY [154] that uses pattern matching of DB query languages as a mapping algorithm. SAVVY does not apply any NLQ interpretation modules or parsing elaborations for the mapping process. Hence, its translation accuracy and overall performance is highly jeopardized.

SQL Formation vs SQL Templates

The current research work simplifies SQL queries generation by using ready SQL templates. Yet SQL construction constraints are used in the mapping algorithm to guarantee accurate SQL template selection. Despite the presence of flexible SQL templates, some recent works [52, 91, 105, 107, 108, 120] still use other methods to construct their own SQLs from scratch, which adds an extra computational complexity to the language translation system. An example is [91], which defined a concept-based query language to facilitate SQL construction by means of NLQ Conceptual Abstraction [88]. This work adds an additional unnecessary complex layer on top of the original system architecture.

Furthermore, [103, 106] used semantic grammar analysis to store all grammatical words to be used for mapping NLQ's intermediate semantic representation into SQLs. Due to this system's complexity, this architecture can only translate simple NLQs, but not flexible with nested or cascaded SQLs. In comparison, the current proposed system does not map whole NLQs into existing SQLs, but maps NLQ lexica to the SQL clauses and keywords. This is to enable the translator algorithm to be domain-independent and configurable on any other environment, without the need of developing a training and testing datasets of NLQ/SQL pairs for every new domain such as in [24].

In [24], a dataset of labeled NLQ/SQL pairs training and testing datasets are generated and classified to correct or incorrect using Kernel Functions and an SVM classifier. This mapping algorithm is at the syntactic level using NLQ semantics to build syntactic trees to select SQLs according to their probability scores. Giordani and Moschitti [24] applies the statistical and shallow Charniak's Syntactic Parser [126] to compute the number of shared high-level semantics and common syntactic substructures between two trees and produce the union of the shallow feature spaces [24]. Such exclusive domain-specific systems are highly expensive and their performance is subjective to the accuracy and correctness of the employed training and testing datasets, which are manually written by a human domain expert. As such, the KRISP system [78] achieved a 78% recall of correctly retrieved SQL answers, while the current research work achieved a 96% recall on the small RDB (2.5 MB) and a 93% recall on the large one (200.5 MB) due to the use of a light weighted mapping algorithm mapped to ready SQL templates.

Another example of recent works that generate their own SQLs is [115], which used syntactic trees of NLQ/SQL pairs as an SQL compiler to derive NLQ parsing trees. In [115], NLQ tokens' lexical dependencies, DB schema and some synonym relations are used to map DB lexica with the SQL clauses via a Heuristic Weighting Scheme [41]. Because [115] does not use any NLQ annotated meaning resources (i.e., Prolog data [155] or Lambda Calculus [129]) or any other manual semantic interpretation and representation to fully understand the NLQ, the SQL generator performance was considerably low. Therefore, authors of [115] applied a reranker [120] to try and boost accuracy using an advanced Heuristic Weights-Based SVM-Ranker [36] based on String Tree Kernels [128]. The reranker indeed increased the recall of correct answers up to 94%, which is still lower than the recall of the proposed research work. This is because in the current work, RDB lexica MetaTable is used for lexical relations disambiguation. A mapping table is also used, which includes RDB data types, PKs and FKs and names of entities (unique values), in addition to other rule-based mapping constraints. Hence, building an SQL generator is more complicated in the language translation field and as a result increases the complexity of the translation algorithm. This is the main reason the current research work uses SQL templates and puts extra focus on passing accurate RDB lexia into SQL templates generator for a better performance and output.

Neural networks have not been used in the current research work, nor for any of the mapping mechanisms. The reason why will be clearer with some recent work examples such as [124, 156]. SEQ2SQL [124] is a deep Sequence to Sequence Neural Network Algorithm [157] for generating an SQL from an NLQ semantic parsing tree. SEQ2SQL uses Reinforcement Learning Algorithm [124] and rewards from in-the-loop query execution to learn an SQL generation policy. It uses a dataset of 80,654 hand-annotated NLQ/SQL pairs to generate the SQL conditions which is incompatible with Cross Entropy Loss Optimization [158] training tasks. This Seq2SQL execution accuracy is 59.4% and the logical form accuracy is 48.3%.

SEQ2SQL does not use any manually written rule-based grammar like what is implemented in the current research work. In another recent work in 2019 [156], a sequence-to-sequence neural network model has been proved to be inefficient and unscalable on large RDBs. Moreover, SQLNet [124] is a mapping algorithm without the use of a reinforcement learning algorithm. SQLNet showed small improvements only by training an MLA sequence-to-sequence-style model to generate SQL queries when order does not matter as a solution to the "order-matters" problem. Xu et al. [124] used Dependency Graphs [116] and the Column Attention Mechanism [159] for performance improvement. Though this work combined most novel techniques, the model has to be frequently and periodically retrained to reflect the latest dataset updates, which increases the system's maintenance costs and computational complexity.

The work in [157] overcomes the shortcomings of sequence-to-sequence models through a Deep-Learning-Based Model [124] for SQL generation by predicting and generating the SQL directly for any given NLQ. Then, the model edits the SQL with the Attentive-Copying Mechanism [160], a Recover Technique [3] and Task-Specific Look-Up Tables [161]. Though this recent work proved its flexibility and efficiency, the authors had to create their own NLQ/SQL pairs manually. Besides, they also had to customize the used RDB, which is a kind of over-customization to the used framework and environment applied on. Hence, results are highly questionable in terms of generalizability, applicability and adaptability on other domains. On the other hand, the current research work used RDBs that are public sources namely, Zomato and WikiSQL.

Intermediate Representations

The current research work tries to save every possible information given by the NLQ tokens so that each of them is used and represented in the SQL clauses and expressions production. Therefore, multiple NLP tools, MetaTables and mapping tables (for unique values) are implemented to fully understand the NLQ and map its tokens to their corresponding RDB elements. Then, the identified attributes are fed into the SQL SELECT clause, while the tables are extracted from the SELECT clause to generate an SQL FROM clause, and the values are used as conditional statements in the WHERE clause. For this simple mapping purpose, other works in the literature use additional intermediate layers to represent NLQ tokens as SQL clauses, which, upon investigation, turned to be not as effective as the NLP tools, MetaTables and mapping tables.

An example of the NLQ intermediate semantic representation layers is using Regular Expressions (regexps) [101] to represent NLP tokens phonology and morphology. This representation happens by applying First Order Predicate Calculus Logic [162] using DB-Oriented Logical Form (DBLF) and Conceptual Logical Form (CLF) with some SQL operators and functions to build and generate SQLs [107]. Yet, the use of regular expressions "regexps" collections in NLQ sentences are not clearly articulated in the literature.

Another example is CliniDAL [27], which used EAV type of DB metadata and grammatical parse trees to process NLQ tokens to then be mapped to their internal conceptual representation layer using the Similarity-Based Top-K Algorithm [138]. More examples of the intermediate layers include PHLIQA1 [4], ASK [46], USL [109], and SEMRESP [7, 129, 111, 122] which defined intermediate tokens meaning representations manually.

Also, the supervision-extensive system [7, 113, 126] used Lambda Calculus to map tokens to their corresponding meaning representations. TEAM [16] adopted intermediate semantic representation layers connected with a DB conceptual schema. In addition, L2S [92] transforms DB lexica into an intermediate tree-like graph then extracts the SQL from the Maximum Bipartite Matching Algorithm [163].

All those great efforts unfortunately proved to be not as effective because of the high complexity and the time-consuming nature of the approaches. Those deficiencies mandated the introduction of a new system that translates NLQs into SQLs while maintaining a high simplicity and performance presented in the current research. The proposed research highlights a new solution to NLP and language translation problems.

Table 17 (Appendix 9) highlights the main similar works in the literature with their advantages and disadvantages summarized in comparison with the proposed work in the current research.

In what follows, a layout of the similar works presented in a chronological order. This section is also summarized in Table 17, Appendix 9 (Figure 5).

NLQ into SQL mapping Approaches
- Authoring Interface Based Systems
- Enriching the NLQ/SQL Pairs via Inductive Logic Programming
- Using MLA Algorithms
- Restricted NLQ Input
- Lambda Calculus
- Tree Kernels Models
- Unified Modeling Language (UML)
- Weighted Links

NLQ Tokens into RDB Lexica Mapping (NLQ Tokens Extraction)
- Morphological and Word Group Analyzers
- Pattern Matching
- Name Entity Recognizer (NER) Alone with Coltech-Parser in GATE
- Java Annotation Patterns Engine (JAPE) Grammars
- Porter Algorithm
- Unification-Based Learning (UBL) Algorithm
- Dependency Syntactic Parsing
- Separate Value and Table Extractor Interfaces

NLQ Tokens into RDB Lexica Mapping (RDB Lexica Mapping)
- Spider System
- WordNet alone

NLQ Tokens into RDB Lexica Mapping (RDB Lexica Relationships)
- Stanford Dependencies Parser
- Dependency Syntactic Parsing
- Dependency-Based Compositional Semantics (DCS) System Enriched with Prototype Triggers

NLQ Tokens into RDB Lexica Mapping (NLP syntax and semantics)
- Named Entity Tagger
- Dependency Parser
- LIFER/LADDER Method
- NLQ/SQL Syntax Trees Encoded Via Kernel Functions
- The Probabilistic Context Free Grammar (PCFG) Method

RDB Lexica into SQL Clauses Mapping (SQL clauses mapping)
- The Extended UML Class Diagrams Representations
- RDB Relationships and Linguistic Analysis

RDB Lexica into SQL Clauses Mapping (Complexity vs Performance)
- L2S System
- Bipartite Tree-Like Graph-Based Processing Model
- Ellipsis Method
- The Highest Possibility Selection
- Weighted Neural Networks and Stanford Dependencies Collapsed (SDC)
- Pattern Matching of SQL

RDB Lexica into SQL Clauses Mapping (SQL Formation vs SQL Templates)
- NLQ Conceptual Abstraction
- Semantic Grammar Analysis
- Kernel Functions, SVM Classifier, and the Statistical and Shallow Charniak's Syntactic Parser
- Heuristic Weighting Scheme
- A Deep Sequence to Sequence Neural Network
- MLA Sequence-To-Sequence-Style Model
- A Deep-Learning-Based Model

RDB Lexica into SQL Clauses Mapping (Intermediate Representation)
- Regular Expressions (regexps)
- The Similarity-Based Top-K Algorithm
- Lambda-Calculus
- An Intermediate Tree-Like Graph

FIGURE 5   Research ideas with their current existing solutions.

Implementation Plan

4

The current research framework components start with the NLIDB interface that the user uses to enter the NLQ sentence. To understand the input NLQ, the NLQ must go through POS recognition via the underlying NLP tasks, such as lemmatizing, tokenizing, annotating and rule-based parsing. The following step, disambiguation, is a conditional step that the NLQ will go through only in the case that there was a vague POS word (i.e., Is "content" a noun or an adjective? Is "separate" a verb or an adjective?). After that, the identified NLQ tokens will be delivered to the matcher/mapper step for mapping the tokens into the elements of the RDB schema MetaTables and the identified lexica into the SQL clauses. The matching lexica will be used in the SQL generation step, which will be executed next.

NLQ INPUT INTERFACE

Before running any script, required dependencies and packages, which are all open source and available, must be downloaded and imported through the Python terminal. Then, the user will insert an NLQ into the data input interface. The user will be returned either the generated SQL results from the MySQL DB or an error alert. The error alert could be concerning the entered NLQ linguistic issues or an error of an existence of more than 1 match or no match at all to the NLQ arguments in the RDB MetaTable.

The MetaTables of NLQ and RDB are created by adding span tags to the RDB elements or the NLQ tokens to attach them with their syntactic and semantic roles. They are also annotated with their synonyms using the WordNet ontology functions. The cost of adding the MetaTables data are fractional to the original size of the RDB itself. For small RDBs, it could add extra 3% on top of the original RDB size. For larger RDBs, such as WikiSQL, it could add up to 10% extra storage space. When the RDB changes, the translation processing

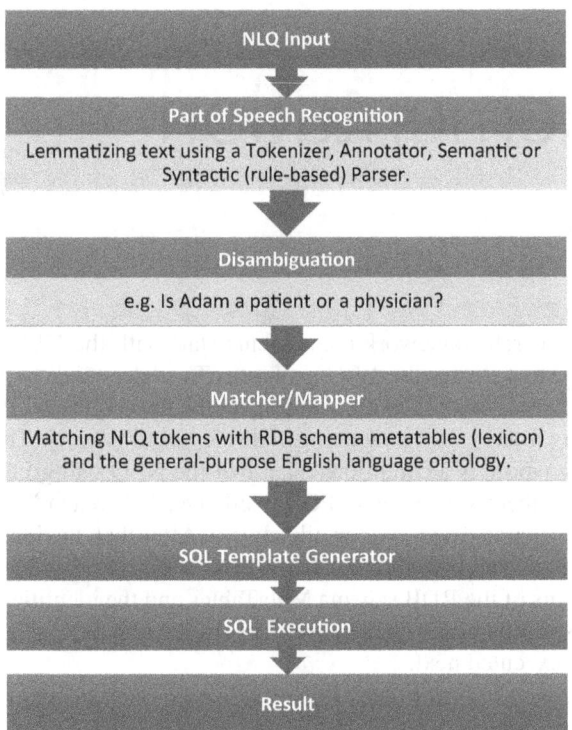

FIGURE 6   Detailed research architecture pipeline.

is not affected by the update because every added record will be automatically annotated by the framework to include necessary annotations and metadata.

The NLQ is inserted through an NLI screen as input data up to 200 characters with the help of two Python libraries, namely, "server.bot", which accepts the input NLQ, and "text_processing.text_nlp", which initially processes the NLQ by separating the words and passing them as arguments to the next module. The NLI will identify NLQ words as arguments, which will later help preparing them for identifying their semantic and syntactic roles. Figure 6 briefly summarizes the steps taken to transform an NLQ into an SQL statement. Those steps will be further clarified throughout this chapter.

POS RECOGNITION

The multilayered translation algorithm framework splits the NLQ into its constituent tokens. Then, these tokens are compared with the RDB MetaTables' contents to single out keywords in the NLQ sentence. With the tokens matching schema data, a.k.a. the lexica, the NLQ should be able to be parsed semantically to identify tokens' semantic-role frames (i.e., noun, verb, etc.) which helps the translation process. Semantic parsing is done by generating the parsing tree using the Stanford CoreNLP library, with input from the English language ontology, WordNet, which feeds the system with NLQ words meanings (semantics).

The first process performed on the NLQ string is lemmatizing and stemming its words into their broken-down original root forms. This is done by deleting the words' inflectional endings and returning them to their base forms, such as transforming 'entries' into 'entry'. Lemmatizing eases the selection and mapping process of equivalent RDB elements. It also facilitates the tokens' syntactic and semantic meaning recognition. Then comes the steps of parsing and tokenizing the words' stems into tokens according to the predefined grammatical rules and the built-in syntactic roles. Those syntactic roles will be mapped to specific RDB elements, for instance, NLQ verbs are mapped with RDB relationships.

PSEUDOCODE 1 ALGORITHM TO CONSTRUCT AN SQL QUERY FROM AN NLQ INPUT

```
Begin
    Split NLQ text to individual ordered words and store
into string array A
    Delete any escape words from A
    Map words in array A with RDB elements E
    Replace words in array A by their matching synonyms
and type from E
    If there is ambiguate word W in A then
        Ask user "What is W?" and match word W with E
    End If
    If there is a conditional phrase C in A
        Replace C with equivalent conditional operator
in O
        Attach O to conditioned attribute name as a
suffix and store in A
    End If
    Do
```

```
         Store attributes and their conditional
operators and tables and relationships for matched
elements E in array R
         Generate SQL template matching the number and
type of tokens in R
         Construct SQL query using array R tokens and
store it in variable Q
    While for each table or attribute or relationship or
conditional operator in array R
         Execute generated SQL query
End
```

For any NLQ translation process, both the parsed tokens and their subsequent POS tags must be clearly and accurately identified. This is performed by an underlying multilayered pipeline which starts with tagging an NLQ POS. Then, the tokenizer, annotator, semantic and syntactic (rule-based) parsers will be applied and any punctuation marks will be removed. Part of this step is omitting the meaningless excess escape words that are predefined in the system (i.e., a, an, to, of, in, at, are, whose, for, etc.) from the NLQ words group. After parsing, a parse tree is generated and a dictionary of tokens' names, syntactic roles and synonyms are maintained in the NLQ MetaTable. Also, the NLQ's subjects, objects, verbs and other linguistic roles are identified. Hence, each tokenized word is registered into the NLQ MetaTable by the syntactic analyzer. Tokens are then passed to the semantic analyzer for further processing.

The semantic analyzer employs a word-type identifier using a language vocabulary dictionary or ontology such as WordNet. The word-type identifier, such as WordNet, identifies what semantic role does a word or a phrase (i.e., common or proper noun) play in a sentence and what is their role assigner (the verb). Furthermore, the semantic analyzer is able to identify conditional or symbolic words and map them with their relative representation from the language ontology. For example, the phrase "bigger than" will be replaced by the operator ">". In other words, the semantic analyzer's entity annotator detects the conditional or symbolic words amongst the input NLQ entities. Then, it replaces them with their equivalent semantic types identified previously by the schema annotator. The entities replacement creates a new form of the same NLQ that is easier for the SQL generator or pattern matcher to detect.

The entity annotator is not the only annotator the NLQ deals with. There are other annotators the NLQ gets passed through such as the numerical annotator, date annotator, comparator annotator, etc. In future work, this step can be further improved to search for the previous annotation results to check for any stored matching patterns of lexicalized rules. This step shall help determine the suited SQL template type or its further sub-queries' divisions faster.

The NLQ gets converted into a stream of tokens and a token ID is provided to each word of the NLQ. The tokens are classified into their linguistic categories such as nouns, pronouns, verbs or literal values (string/integer variables). The algorithm maps the tokens into tables, attributes, values or relationships according to their linguistic categories and semantic roles. The rest of the acquired information will be used to formulate SQL query clauses (i.e., comparative or operational expressions) according to the tagged tokens.

The Python NLP lightweight library (TextBlob) is used as an NLQ POS recognizer (i.e., "speech_recognition" library). NLQ tagger and lemmatizer are implemented to facilitate the equivalent RDB elements selection. In addition, other libraries are also considered including, but not limited to, "string_punctuation", "Stanford CoreNLPspellcheck", "nltk.corpus" (using WordNet), "textblob.word", "wordNetLemmatizer", "nltk.stem", "sentence_tokenize", "word_tokenize", "nltk.tokenize", "unicodedata" (for mathematical operations and symbols), "textt_processing", "text_nlp" and "server.tokenizer". Those libraries' usage and distribution is explained in Figures 7 and 16 (Chapter 6).

The system checks each NLQ word's semantic role and adds it to the registry to be passed on to the next step, as illustrated in PseudoCode 2 (Appendix 1). For example, if the first element in the list (index[0]) is a common noun, the code would check if the NLQ word is a table. Also, if there is a corresponding attribute to a value, add the word to the values list, and so on. The algorithm applies the 'Maximum Length Algorithm', illustrated in PseudoCode 3, to remove tokens from the attributes list if the tokens are also in the values list. This algorithm enables the system to avoid duplicate use of the same tokens, which helps in avoiding potential errors and inaccuracy.

PSEUDOCODE 3 MAXIMUM LENGTH ALGORITHM

```
for values(a, v)
    if a ∈ attributes()
        remove a from attributes()
    end if
end for
```

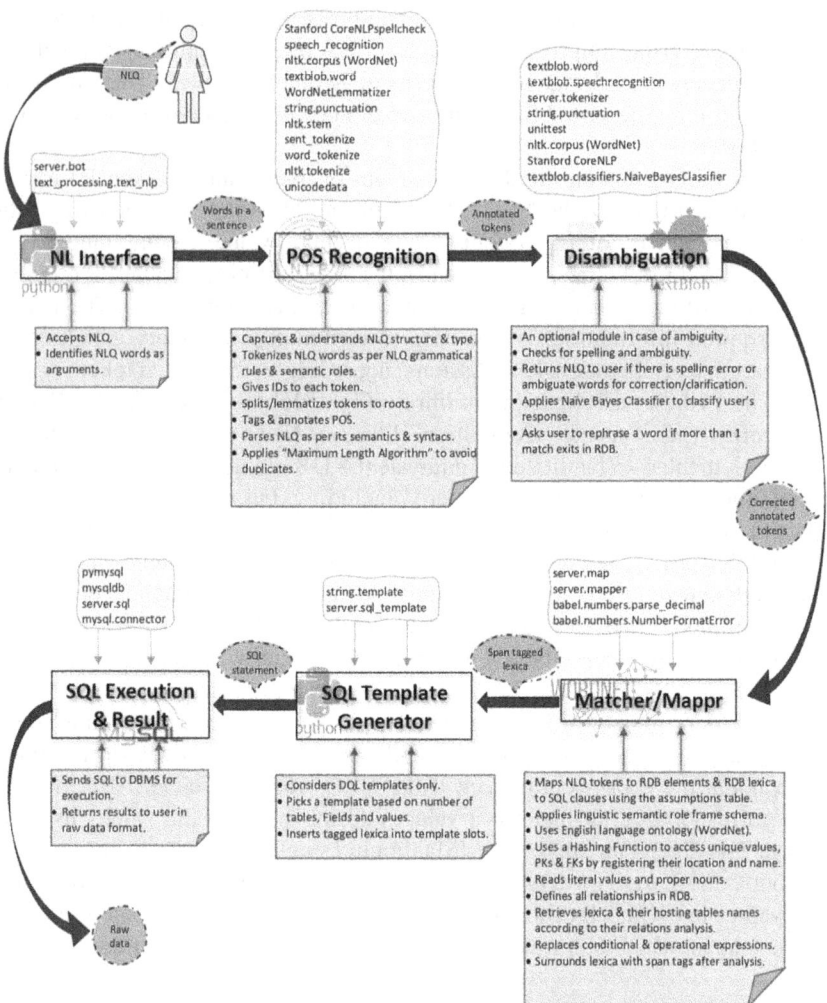

FIGURE 7 Detailed research organization pipeline (light gray boxes are Python libraries; dark gray boxes are tasks & gray blobs are passed-on data).

To recognize literal values, the functions "parse_decimal" and "Number FormatError" are used from the "babel.numbers" library. "parser" and "WordNetLemmatizer" functions from the "nltk.stem" Python library are used to insert the RDB elements' synonyms into the RDB schema automatically.

This happens by adding the synonym and then creating the relationship 'IS_LIKE' with the corresponding RDB element as illustrated in PseudoCode 4.

PSEUDOCODE 4 SYNONYMS MATCHING WITH RDB ELEMENTS

```
insert_synonyms()
    for s ∈ synonyms and e ∈ elements
        if s is similar to e and similarity > 0.75 then
            merge (s, e) as (s)-[IS_LIKE]->(e)
        end if
    end for
```

During the tokenizer and tagger stage, illustrated in PseudoCode 5 (Appendix 2), NLQ POS, semantic analysis and syntactic representation are expressed and extracted by tagging them. This helps to extract the tokens' linguistic sub-parts (i.e., adjectives and noun/verb phrases) for accurate mapping later. The POS tagger will tag each NLQ token to define its syntactic role. The tokenizer analyzes the NLQ token types discussed in Table 4 and returns the extracted sentence structure parts (i.e., verbs, noun phrases) mentioned in the NLQ, together with the tagged version of the lemmatized tokens. In addition, tokens' semantic roles under the SQL scope, which is a value, attribute, table or a relationship, will also be returned.

Not only RDB elements are tagged with their synonyms, but SQL keywords are also tagged with synonyms and semantic information. This tagging happens using the semantics dictionary (WordNet) and the "nltk.tokenize" libraries, namely, "sentence_tokenize" and "word_tokenize", as illustrated in PseudoCode 6.

PSEUDOCODE 6 SQL KEYWORDS TAGGING WITH THEIR SYNONYMS

```
sql_tagging()
    for attributes and tables and conditions
        if sql ≠ ∅ then
            apply semantics_dict[synonyms]
        add attributes synonyms to select
        add tables synonyms to from
        add conditions synonyms to where
        end if
    end for
return sql_tagging(tags)
```

Other SQL keywords such as aggregate functions (e.g., AVG, SUM, etc.) or comparison operations (e.g., >, <, =, etc.), defined in the Python "unicode-data" library, are also tagged with their synonyms for easy and accurate mapping, as illustrated in PseudoCode 7 (Appendix 3).

After all of the NLQ and SQL words are tagged with their synonyms, the algorithm will start the testing module to validate the similarity of RDB Lexica and NLQ tokens compared with their tagged synonyms. If the "Similarity" is greater than or equal to 75% (the least-acceptable similarity variance), it is considered a matching synonym. In this case, lexica or tokens are tagged with their matching synonyms according to their semantics using WordNet synonym datasets.

DISAMBIGUATION

NLQ input disambiguation is an intermediate process and is done through contextual analysis. When the system cannot make a decision due to some ambiguity, it asks the user for further input. This occurs in case of the presence of more than one match for a particular NLQ token (e.g., "Is Adam a patient or a physician?"). However, engaging the user is solely for clarifying a certain ambiguity in the NLQ input by choosing from a list of suggestions of similar words or synonyms present in the lexica list.

In future work, and as a further disambiguation step to guarantee generated SQL accuracy, a feedback system could be applied after NLQ analysis. This feedback system asks the user to confirm the translated NLQ into SQL query by asking the user "is this the desired SQL?". However, since we assume the user's ignorance of any programming abilities, including SQL, this feedback system is not applied in the current research work.

The RDB elements with identical names are carefully managed according to the NLQ MetaTable (Table 1) and the RDB elements' MetaTable (Table 2). Hence, the ambiguity-checking module will eventually have a list of all identically named elements and their locations in the RDB.

Every entered NLQ goes through a syntactic rules checker for any grammatical mistakes. This module checks the NLQ validity or the need for a user clarification for any ambiguity or spelling mistakes using the Python libraries "unittest" and "textblob". The algorithm will proceed to the next step if the NLQ is valid. Otherwise, the algorithm will look for a clarification or spelling correction response from the user by asking them to choose from a few potential corrections. Then, the user's response is classified to either positive (i.e., Yes) or Negative (i.e., No). This classification happens using the Naïve

Bayes Classifier from the prebuilt Python library "textblob.classifiers" used as an SQL grammar classifier. If the user's response is positive, it will use the corrected form, otherwise, it will use their original NLQ form and work with it. This step uses the "Stanford CoreNLP" and "nltk.corpus" libraries to check for the NLQ validity and uses the syntactic rules checker to check for spelling errors as illustrated in PseudoCode 8 (Appendix 4).

MATCHER/MAPPER

In this phase, synonyms of NLQ tokens are replaced with their equivalent names from the embedded lexica list. Then, SQL keywords are mapped and appended with their corresponding RDB lexica. The Matcher/Mapper module applies all mapping conditions listed later in Table 4 which covers NLQ tokens, their associated RDB lexica, SQL clauses, conditional or operational expressions or mathematical symbols.

This module has access to MetaTables (data dictionaries) of all attributes, relationships, tables and unique values (Mapping Tables). Both mappers in Figure 8 can refer to an embedded linguistic semantic-role frame schema, data or language dictionary, or the underlying RDB schema. This layer uses RDB schema knowledge (the semantic data models, MetaTables) and related syntactic knowledge to properly map NLQ tokens to the related RDB structure and contents.

In regard to unique RDB values, and since it's a storage crisis to store all RDB values in a RAM or CACHE memory, only unique values and PKs and FKs will be stored in a mapping table. The unique values' hosting attributes and tables will be specified, and a hashing function will be used to access them.

For smaller RDBs (i.e., Zomato), and, as explained in Table 3, the mapping table is built using the Python dictionary "server.map" that finds associations

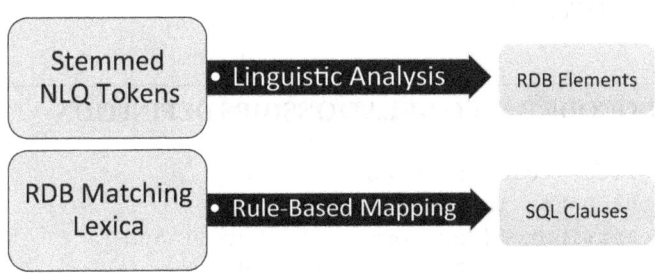

FIGURE 8 The two embedded mappers.

TABLE 3   Mapping table design options

RDB SIZE	EXAMPLE	APPLIED TECHNIQUE	DATA USED
Small	Zomato	Python dictionary "server.map"	Associations between NLQ tokens and RDB elements
Large	WikiSQL	Data profiling	RDB elements' statistics

between NLQ tokens and RDB elements that are often queried together. For larger RDBs (i.e., WikiSQL), data profiling is performed on RDB elements' statistics to automatically compile the mapping table. This compilation is based on which RDB elements are queried more often, and then stored in the mapping table as a hashing function. The mapping table is expressed as mapping_table[unique_value] = corresponding_attribute.

Compared to the great value the mapping table adds to the algorithm's accuracy, there would not be any significant overhead added by integrating a mapping table. Ye, the bigger the RDB the bigger the mapping table size, which affects resources usage in terms of storage capacity.

Mapping NLQ Tokens into RDB Elements

This unit matches NLQ tokens with RDB schema MetaTables (lexica list) to check for their existence. This unit also checks the general-purpose English language ontology (WordNet) for NLQ tokens' synonyms and meanings. Before discussing the mapping algorithm itself, the RDB schema relationships, its lexica (tables' and attributes' names) and the conditional and operational expressions must be defined.

First, relationships in the RDB schema will be defined and registered. Then, they will be matched with the NLQ verb. If the NLQ verb is unidentified or missing, the relationship between the NLQ tokens will be found through analysing the lexica intra-relationships with each other as explained step by step in PseudoCode 9.

PSEUDOCODE 9   NLQ RELATIONSHIPS DEFINITION

```
/* register relationships */
for attributes in rdbSchema do
  for attribute1(lexicon1, attribute1) and
attribute2(lexicon2, attribute2) do
      relationships ← relation(attribute1, attribute2)
```

```
   end for
end for
/* if NLQ has no verbs */
if nlq(verb) = True
   check relationships(synonyms)
else
  check relation(attributes) in relationships
end if
```

Now that RDB relationships have been defined and registered, the algorithm is able to retrieve matching RDB lexica, and their hosting attribute or table. This matching happens in accordance with the matching NLQ lexica and the relationships built between them. The retrieved data will be then passed on to the next step to be used in the SQL clauses mapping as explained in PseudoCode 10.

**PSEUDOCODE 10 FINDING RDB LEXICON PARENT
ATTRIBUTE, TABLE AND RELATIONSHIPS**

```
for rdbSchema(lexicon) do
    find parent and relationship
  return parent(attribute), parent(table),
Relationship(verb)
```

NLQ tokens are mapped with their internal representation in the RDB schema via the MetaTables and synonyms, and then mapped to the SQL clauses. Each input token is mapped with its associated RDB element (lexicon) category (e.g., value, column, table or relationship).

The mapper translates the NLQ literal conditions and constraints, whether they are temporal or event-based, into the SQL query clauses such as translating "Older than 30" to "Age > 30". The mapper also extracts matches of function or structure words (i.e., linking words or comparison words) and search tokens (i.e., Wh-question words) from the annotated NLQ. Function words could be prepositions (i.e., of, in, between, at), pronouns (i.e., he, they, it), determiners (i.e., the, a, my, neither), conjunctions (i.e., and, or, when, while), auxiliary (i.e., is, am, are, have, got) or particles (i.e., as, no, not).

This module checks for the presence of any NLQ conditional, operational or mathematical expressions (i.e., min, max, avg, etc.) in the NLQ to customize the WHERE statement accordingly to retrieve only relevant data from the RDB, as explained in PseudoCode 11.

PSEUDOCODE 11   CHECKING NLQ FOR EXPRESSIONS

```
if nlq(words) ← expr(cond, oper, math)
    where_clause = True
    adjust where_clause with conditional[] or
operational[] or mathematical[]
else
    where_clause = False
end if
```

TABLE 4   Main rule-based assumptions

NLQ TOKEN TYPE	EXAMPLE	SCHEMA CATEGORY	SQL SLOT
Instance (Proper Noun)	Sarah	Value	WHERE condition
Adjective, Adverb, Gerund	Strongly	Attribute	SELECT or WHERE (if accompanied with a value) clause
Number (Literal Value)	100	Value	WHERE condition
Common Noun	Patient	Attribute/Table	SELECT/FROM selection operator clause
Comparative Expression	Most	Conditional Values	MAX, MIN, AVG, etc. clauses or with WHERE clause
Comparative Operation	Equal	Conditional Values	=, >, <, <>, ><, >=, <=, etc. with WHERE clause
Verb	has	Relationship	WHERE condition, JOIN, AS, or IN
Wh-phrases	What	Value's Attribute Indicator	N/A
Prepositions	With	N/A	N/A
Conjunction/ Disjunction	And	N/A	WHERE condition AND, OR, etc.

During the parsing phase, the NLQ is decomposed into a head-noun, noun modifiers, verbs that relate semantic roles together, objects and relationships descriptivism and adjectives or adverbs that describe verbs. Hence, NLQ tokens can be any of the token types in Table 4.

According to those observation-based assumptions summarized in Table 4, an SQL template can be easily generated. The SQL template generator mainly needs to know the number of attributes, tables and relationships. In addition, further information are fed to the SQL generator, such as the AND/OR clauses (for JOIN clauses), conditional comparative expressions (for WHERE or AGGREGATE clauses), the conditional comparative operations (for INTERVAL clauses or controlled values) and numbers and instances (literals) as values. For example, if the token is a value, then the corresponding attribute (object) is its column name.

Moreover, synonyms of SQL clauses are also considered. For example, 'search', 'show', 'find', 'get' or even the word 'select' are all synonyms of the SQL clause "SELECT". Similarly, 'count', 'how many' or 'how much' are synonyms of the "COUNT" statement. Also, 'where', 'who has' or 'with' are synonyms of the "WHERE" clause.

Since the RDB lexica may not be explicitly used in the NLQ, the matcher/mapper unit tries to match an NLQ token to an equivalent RDB lexicon by comparing every token (and its synonyms) to its potential RDB element (or its synonyms). In addition, NLQ verbs will also be matched with their equivalent RDB schema relationships to locate where in the RDB schema is this token being referenced to. If a match is found, the algorithm replaces the token with the matching lexicon and returns the match in the form (table, attribute, value, relationship) with each element surrounded by span tags. If the token is found to be an RDB value, the attribute and subsequent table will be known automatically. This step uses the Python library "server.mapper" as explained in PseudoCode 12 (Appendix 5).

Dependency trees, derived from the Stanford CoreNLP syntactic trees, are used to explain the relationships between any two values based on their connectivity status or having common parent node in the dependency tree.

RDBs illustrate relationships based on the data types. As such, values and attributes existing in the same entity (column) are related, so as attributes' tables or tables connected with a particular relationship.

In the current research work, verbs will be mapped to associated relationships, and the verbs' parameters (subject or object noun phrases) into their corresponding attributes in the lexica list. Identifying this relationship association proved to increase the corresponding attributes selection accuracy. Thus, the tokens list is defined in this phase by the lexical analyzer using the language ontology WordNet, and eventually replaced by the RDB MetaTable lexica and passed to the syntactic analyzer.

Mapping RDB Lexica into SQL Clauses

This mapping uses the proposed rule-based algorithm that is based on the assumptions table (Table 4).

The first step is building the main SQL clauses, the SELECT, FROM and WHERE clauses. The attribute names will be fed into the SELECT clause. Hence, the SELECT keyword is appended with the table attributes. Attributes are identified by semantically analysing the Wh-word's main noun phrase or head noun (main noun in a noun phrase). The WHERE keyword is mapped with the attribute-value pairs derived from the NLQ semantics. The FROM keyword is mapped with all involved tables' names referenced in the SELECT and WHERE clauses. If there is more than one table, tables will be joined and added to the FROM clause. If there is a data retrieval condition, a WHERE clause will be added, and conditions will be joined as illustrated in PseudoCode 13 (Appendix 6).

In this phase, the key mapping function is mapping SQL clauses and keywords with the NLQ identified lexica, and then building the SQL query. The tables list which tables names should be selected from must be identified. The list of relationships, attributes and values with their associated attributes should also be identified in the form (attribute, value).

SQL TEMPLATE GENERATOR

SQL formation is done in this stage. SQL components (i.e., tables' names, lexica hosting sources, attribute-value pairs, data retrieval conditions and relationships) are identified from the input NLQ and arranged in a proper sequence. The identified NLQ lexica, schema matching elements, and the identified operators (if any) are then fed into the SQL template generator to generate a proper SQL statement. SQL templates will be selected based on the numbers of identified tables, attributes and attribute-value pairs. After that, the system establishes a connection with the RDB to transfer SQLs to the RDBMS for execution.

The RDB schema contains unique identifiers (e.g., PKs and FKs) list stored in a dependency table (the mapping table). This table uniquely identifies each instance of each attribute, and whether they are connected via a relationship with any other RDB elements. Each attribute's unique identifier is added to the SQL query constraints to guarantee that only the particular information of interest is returned.

The SQL templates list of SELECT statements considers possible SQLs depending on the NLQ question and desired answer using all input from previous steps. For NLQs with explicit SELECT parameters, the proposed assumptions-based system uses an investigative heuristic procedure to determine what parameters belong in which SQL slot. The parameters can be used either as query constraints (e.g., WHERE, IN, etc.) if they already have values, or as part of the SELECT statement; if they need their values to be retrieved from the RDB. After that, the input values and necessary operators are used to construct the query constraints in the proper SQL template. An example of assigning a suitable operator for every WHERE conditional pair (attributes and values) is converting the NLQ string "equal" to the SQL keyword "LIKE" or the operator "=" or converting "smaller or equal" to the operator "<=".

In this work, only the following SQL main clauses are considered, in addition to other supplementary clauses (e.g., AS, COUNT, etc.):

- SELECT: identifies desired attributes to be retrieved according to the NLQ processor.
- FROM: identifies the tables where the SELECT attributes are from, or where the attribute-value pairs appearing in the WHERE conditional statements are originally from. In case of multiple tables, relationships between tables are identified using JOIN.
- WHERE: identifies the conditions and criteria that must be applied on the retrieved data in a form of conjunctions of attributes and their desired values. If there is more than 1 table, JOIN conditions are used.

Furthermore, it is important to determine SQL classes which a system can or cannot generate. All adopted SQL queries are simple, covering SELECT and WHERE clauses. Therefore, the supported SQL statements are declared in Figure 9 (Chapter 5).

The chosen SQL template solely relies on the number of NLQ tokens related to tables, attributes and values. Yet the type of the generated SQL template could be nested, aggregated, negated, or basic selection, joining and projection. Those types are further categorized in Appendix7.

The Python libraries "string.template" and "server.sql_templates" are used to construct and generate SQL statements in the form: SELECT {attributes} FROM {table} [, {table}] (WHERE {attribute=value} [and {attribute=value}]). The algorithm uses default template strings (placeholders) until it receives the selected lexica from the previous steps, particularly from the Matcher/Mapper step. This process performs in accordance with Table 4 tokens mapping rules and as illustrated in PseudoCode 14 (Appendix 8). It is important to note that all SQL templates use the DISTINCT keyword as per the embedded Maximum Length Algorithm explained earlier.

The first step in the SQL template generator module is connecting the SQL templates generator environment to the MySQL server and the MySQL DB session via the "mysql.connector" function. The SQL template generator will choose which template should be chosen to generate the query. This SQL template election is based on the required number of involved tables, attributes and attribute-value pairs derived from the RDB schema using MySQL DB and the Python libraries "pymysql" and "server.sql". After selecting the right template, the algorithm will return the generated SQL statement with the lexica inserted appropriately. Then, the query is pushed forward to the MySQL server for execution on the connected RDB as illustrated in PseudoCode 14.

PSEUDOCODE 14 SQL GENERATION AND EXECUTION

```
if    generated_sql = True then
      execute(sql)
else print(Sorry, there were no results for your query!)
end if
```

SQL EXECUTION AND RESULT

After the generated SQL query execution, data is fetched from the RDB and displayed to the user as raw data. An example of usage is the following query line entered into the Python command line interface.

```
python3 -m nlqsql.main -d zomato/city.sql -j output.json -i
'What is the average size of restaurants with name Burrito?'
```

The output would be:

```
{'select': {'attribute':'size', 'type':'AVG'},
'from'  : {'table':'restaurant'},
'where': {'conditions':[{'attribute':'name',
'operator':'=', 'value':'Burrito'},]},}
```

And the execution result is:

```
-----------------
| AVERAGE (*)     |
-----------------
| 23 Square Feet  |
-----------------
```

Implementation User Case Scenario

5

To match an NLQ to a proper SQL template, NLQ text will be analyzed and tokenized to be matched against the RDB index. The NLQ goes through a full-text search after it has been tokenized, which is different from the common keyword search.

Figure 9 shows the directed RDB chart diagram for the Post-Traumatic Stress Disorder (PTSD) RDB. RDB representation is used here instead of ERD because the RDB relationships are richer in information than ERDs [13]. Tables 5–8 are the RDB tables. Table 9 is the NLQ MetaTable and Table 10 is the PTSD RDB elements MetaTable that stores all the entities in the RDB and their metadata.

All definitions of the RDB entities are stored in Table 10 to describe the tables and attributes. Using Tables 9 and 10, the current automatic mapping algorithm can produce considerably accurate mapping results.

RDB keywords related to different tables and attributes are stored together. Hence, the algorithm is able to map the NLQ tokens to their internal representation of source attributes and tables in the RDB. To reduce ambiguity, the relationships between attributes are controlled in the RDB design to be in the form of verbs only (Figure 10).

Table 10 stores all of the definitions of RDB entities. This describes the tables, attributes, and unique values. Using Table 9, the current automatic mapping algorithm can produce more accurate mapping results since it is able to map the NLQ tokens to their internal representation of source attributes and tables in the RDB. This is because all DB keywords related to different tables and attributes are stored together. To reduce ambiguity, the relationships between attributes are controlled in the RDB design to be only verbs.

DOI: 10.1201/b23367-5

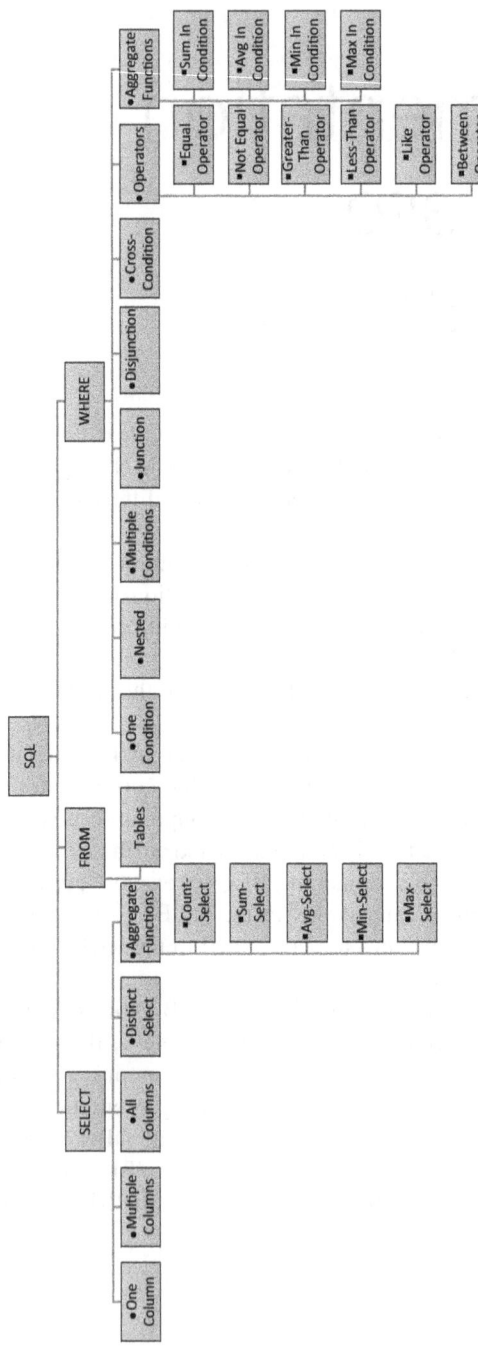

FIGURE 9 Included SQL classes.

TABLE 5 Medications table

P_ID	MED_NAME	MED_CODE
43159	VOLTAREN 75MG TABLET	M01AB05
31896	TYLENOL and CODEINE TAB	N02AA59
32424	ARTHROTEC 50 TABLET	M01AB55
37772	AMITRIPTYLINE HCL 10MG TAB	N06AA09
42235	NASONEX 50 MCG NASAL SPRAY	R01AD09

TABLE 6 Patients table

P_ID	P_SEX	P_BY	P_NAME
43159	Female	1986	Adam
31896	Male	1989	Sarah
32424	Female	1989	Ahmed
37772	Female	1980	Ted
42235	Male	1955	Lin

TABLE 7 Physicians table

PH_ID	PH_NAME	PH_BY
80702	John	1958
80701	Sally	1977
80702	Tom	1980
80703	Matt	1964
80701	Abby	1982

TABLE 8 Diseases table

P_ID	PH_ID	DISEASE_NAME
43159	80702	PTSD
31896	80701	Depressive Disorder
32424	80702	Anxiety Depression Disorder
37772	80703	Hypertension
42235	80701	PTSD

TABLE 9   NLQ MetaTable

WORDS	NATURE	CATEGORY	SYNONYMS
Sarah	Instance	Value	Person, Patient
Has	Verb	Relationship	Own, Obtain, Have
Physician	Noun	Attribute	Doctor, Provider, Psychiatric, Surgeon

TABLE 10   Mapping table for unique values

UNIQUE VALUES	SOURCE	PK/FK	SYNONYMS
Adam	PTSD.Patient.P_Name	No	Person, Patient, ill
43159	PTSD.Patient.P_ID	PK	Patient, Identification, Number
John	PTSD.Physician.Ph_BY	No	Person, Physician, employee
80703	PTSD.Disease.Ph_ID	FK	Physician, Identification, Number
AMITRIPTYLINE HCL 10MG TAB	PTSD.Medication.Med_Name	No	AMITRIPTYLINE, HCL, 10, Milli Gram, Tblet, Drug, Medication

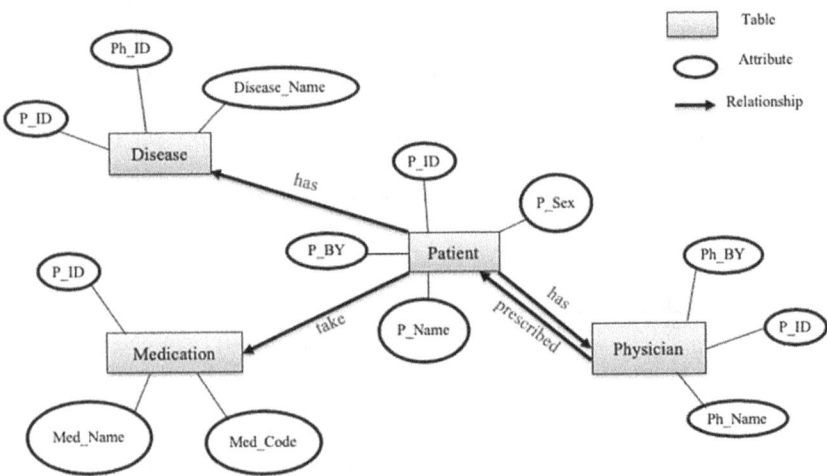

FIGURE 10   PTSD RDB chart diagram.

The examples discussed in the below use case scenarios are as follows:

Q1: What is Adam's birth date? *(Simple Query)*
RA1: Π_{P_BY} ($\sigma_{P\_Name = \text{"Adam"}}$ Patients)
SQL1: SELECT P_BY FROM Patients WHERE P_Name = "Adam";

Q2: Who is the physician that Sarah has? *(Nested Query)*
RA2: $\Pi_{Physician.Ph\_Name}$ ($\sigma_{Patient.P\_Name = \text{"Sarah"}}$ Physician $\bowtie$ Patient)
SQL2: SELECT Physician.Ph_Name FROM Physician INNER JOIN Patient
ON Patient.P_ID = Physician.P_ID WHERE Patient.P_Name = "Sarah";

Q3: What is the most popular illness? *(Simple Query)*
RA3: $\Pi_{Disease\_Name}$ ($\sigma_{MAX (Disease\_Name)}$ Disease)
SQL3: SELECT MAX ([ALL I DISTINCT] Disease_Name) FROM Disease;

Q4: What drug is Ahmed taking? *(Nested Query)*
RA4: $\Pi_{Medication.Med\_Name}$ ($\sigma_{Patient.P\_Name = \text{"Ahmed"}}$ Medication $\bowtie$ Patient)
SQL4: SELECT Medication.Med_Name FROM Medication INNER JOIN
Patient
    ON Patient.P_ID = Medication.P_ID WHERE Patient.P_Name =
"Ahmed";

Q5: What medications did John prescribe for his patients? *(Cascaded Query)*
RA5: $\Pi_{Medication.Med\_Name,\ Medication.Med\_Code}$ ($\sigma_{Physician.Ph\_Name = \text{"John"}}$ (Medication $\bowtie$
Patient) $\bowtie$ Physician)
SQL5: SELECT Med_Name, Med_Code FROM Medication WHERE P_ID
IN
    (SELECT P_ID FROM Patient WHERE P_ID IN
    (SELECT P_ID FROM Physician WHERE Ph_Name = "John")));
Or
    SELECT Medication.Med_Name, Medication.Med_Code FROM
(Medication INNER JOIN Patient
    ON Medication.P_ID = Patient.P_ID) INNER JOIN Physician ON
Physician.Ph_ID = Physician.Ph_ID)
    WHERE Physician.Ph_Name = "John";

USER CASE SCENARIO

Let us assume there is a physician with the below NLQs, how could the proposed algorithm reach the consequent SQL in order to be executed on the system for answers?

Example 1:

Q1: What is Adam's birth date?
The first step is the NLQ words breakdown process to its separate tokens.
Tokens Breakdown:

- Adam = Instance = Value.
- Birth date = noun phrase = attribute.

The NLQ words or phrases considered as tokens are those that present a
particular meaning. Such tokens will eventually participate in the identification of the RDB tables, attributes, relationships, operators (MAX,
AVG) or values. This is because any given token may have 1 of 5 possible
matches: a table, an attribute, a value, a relationship or an operator.

After searching the RDB for the instance "Adam", it was found under
Patients.P_Name. so, the attribute name is found.

The second valuable token is "Birth Date". Since every RDB element (e.g., attribute) has a list of synonyms, BirthDate was matched
with Patients.P_BY. The noun phrase "Birth Date" is also a synonym of
the physician's birth Year (Ph_BY), hence, the system must determine
the best RDB element match among all possible matches. This is done
using knowledge from other tokens' processing. As such, since "Adam"
was found under "Patient" table, then the winning Birth Date match is
"P_BY". Other matches' determining mechanisms involve technical procedures such as statistical similarity measures (e.g., N-Grams Vectors'
Comparison Method). The "P_BY" here will be fed to the WHERE
clause. If there are no WHERE clauses, all DB relations and attributes will
be considered to find valid conditions. Some NLQs might not have conditions, meaning there would not be a WHERE clause in the SQL template.
Generally speaking, any tables mentioned in the SELECT or WHERE
clauses should, by default, be included in the FORM clause to avoid any
SQL execution failure. Efficiency of this approach will be evaluated later
using accuracy measures.

So, the acquired information are:

- Table = Patient
- Attribute 1 = P_BY
- Attribute 2 = P_Name

The SQL Template used here is:

SELECT (Attribute1) FROM (Table) WHERE (Attribute2) = (Value);
Now we have all the information we need to execute the query as follows:

SELECT P_BY FROM Patients WHERE P_Name = "Adam";
Figure 11 summarizes the steps followed to solve example 1.

The benefit from Figure 12, the tokens breakdown analysis diagrams is to show the ability to reach the source attribute, table and related RDB from the NLQ tokens. Finding them helps feeding the SQL template with its necessary arguments.

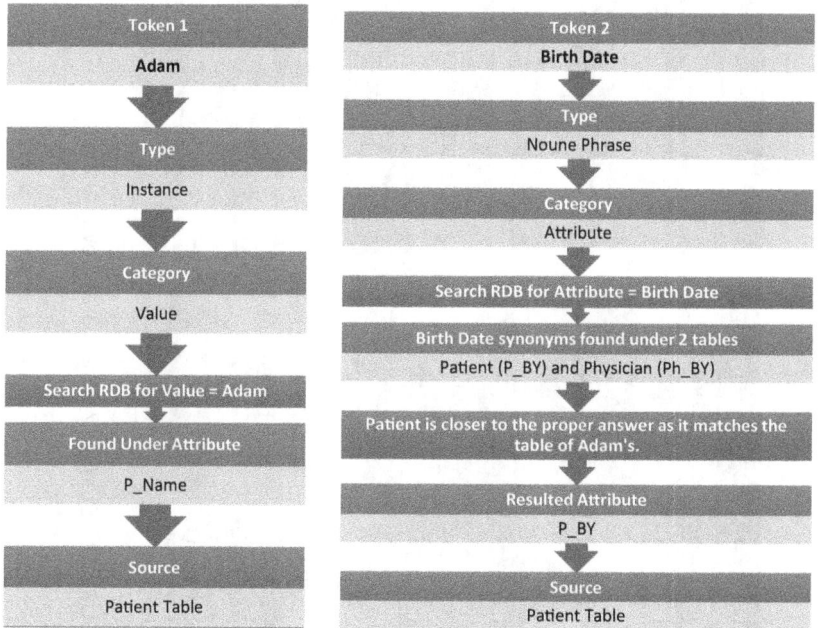

FIGURE 11   Example 1 tokens breakdown analysis.

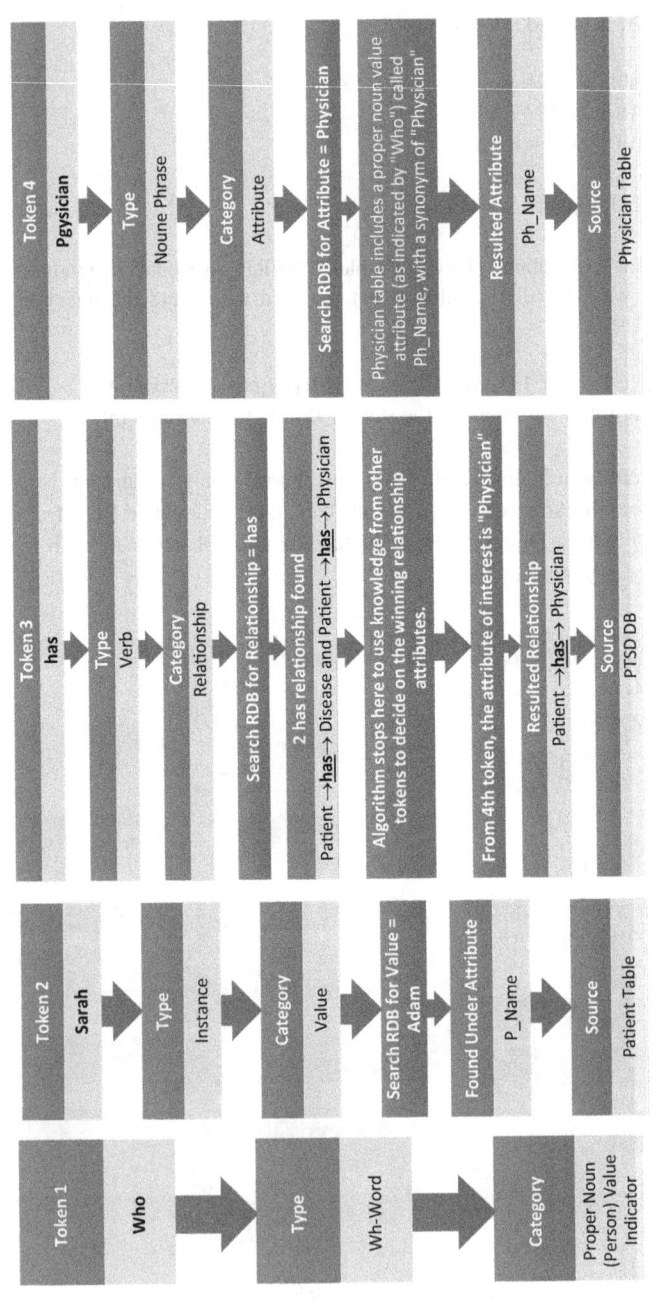

FIGURE 12   Example 2 Tokens Breakdown Analysis.

Example 2:

Q2: Who is the physician that Sarah has?

NLQ Tokens Breakdown:

*   Physician = Common Noun = Attribute
*   Sarah = Instance (Proper Noun) = Value
*   Has = Verb = Relationship

Here, relational DB is necessary because the relationship between the tables "Patient" and "Physician" is important. This relationship will help identify the table that the instance "Sarah" resides in, and who her physician is in the Physician table. This is identified by matching the patient ID in the two tables. All of these processes are computationally expensive in a non-relational DBMS. The difference here is that there is a critical piece of information attached to the relationship between the two tables, which must be a verb, that is translated automatically by the algorithm to match the verb in the NLQ (or its synonyms). Therefore, the acquired information is:

*   Table 1 = Patient
*   Table 2 = Physician
*   Attribute 1 = P_Name
*   Attribute 2 = Ph_Name

For Sarah's identification, it is similar to example 1. This NLQ example has a verb (has), which is translated automatically to a schema relationship. After a search among the schema relationships matching the verb "has" (or its synonyms e.g., have, obtain, acquire, etc.), more than 1 matching relationship appeared coming out of the "Patient" table. Hence, we will use the remaining information we have (Physician) to narrow down the results. The adjacent physician name to the patient name "Sarah" in the "Physician" table will be looked up. This is because this algorithm depends solely on linguistic searching tools, while other matching mechanisms involve technical procedures such as statistical similarity measures (e.g., N-Grams Vectors' Comparison Method).

After searching the RDB for the instance "Sarah", it was found under "Patients.P_Name", so the attribute name (Patients) is automatically found. This attribute name will later be fed to the WHERE clause.

The second valuable token is "Physician". Since every RDB element (e.g., attribute) has a list of synonyms, the table identified here is "Physician", and the Attribute is (Ph_Name) with the synonym "Physician Name".

This attribute was chosen because there is only one table containing the attribute "Physician" as a synonym to the attribute stored in its metadata, "Ph_Name".

Although the word Physician also exists in the Ph_BD attribute metadata, the WH-Word in the NLQ (Who) refers to a human name instance (value), not consecutive digits or a number as in the Physician Birth Date (Ph_BD) attribute values.

Therefore, the acquired information are:

- Table 1 = Patient
- Table 2 = Physician
- Attribute 1 = P_Name
- Attribute 2 = Ph_Name

Since we have more than one table, the suitable SQL Template here is:

SELECT (Table2).(Attribute2) FROM (Table2) INNER JOIN (Table1) ON (Table1).(Attribute3) = (Table2).(Attribute4) WHERE (Table1).(Attribute1) = (Value);

The "Unique Identifiers List" library has a list of all unique IDs. The SQL template will mandate this function to look for the appropriate IDs from both tables to use in filling the SQL template. This will result in identifying P_ID attribute in both tables.

Now that all needed information are found, the algorithm is ready to execute the query with proper join clauses between the two tables as follows:

SELECT Physician.Ph_Name FROM Physician INNER JOIN Patient ON Patient.P_ID = Physician.P_ID WHERE Patient.P_Name = "Sarah";

Example 3:

Q3: What is the most popular illness?
Tokens Breakdown:

- What =Value Indicator
- Most = Comparative Expression = MAX clause
- Illness = Common Noun = Attribute

Based on our main assumptions, any comparative expression will help identify the SQL comparative clause. In this case it is a MAX, and to calculate the maximum of any range, we have to know the values within that range.

From the token "Disease", and after a search around the RDB, we found only one table called disease with a synonym of illness. Under that table, there is one attribute containing the word disease, which is Disease_ Name. This concludes all the required information to use the following SQL template:

SELECT MAX(COUNT(Attribute)) FROM Table;
And with the following acquired information:

- Table = Disease
- Attribute = Disease_Name

Using the identified arguments, we can execute the following SQL query:

SELECT MAX ([ALL | DISTINCT] Disease_Name) FROM Disease;
Figure 13 summarizes the steps taken to solve Example 3.

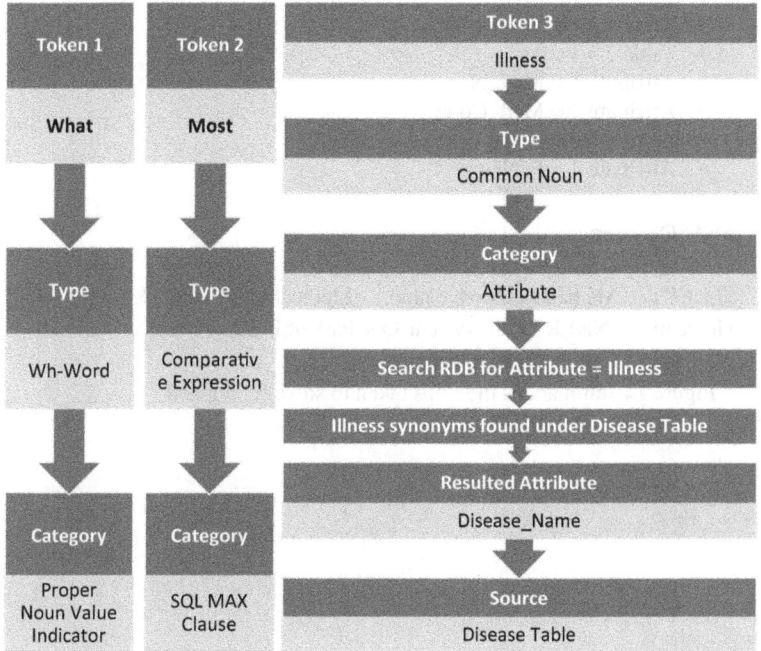

FIGURE 13   Example 3 Tokens Breakdown Analysis.

Example 4:

Q4: What drug is Ahmed taking?
   Tokens Breakdown:

- What=Value Indicator
- Ahmed= Instance = Value
- Drug = Common Noun = Attribute
- Taking = Verb = Relationship

Same as previous examples, except that the word "drug" has more than 1 matching. We have 1 table called Medication with a synonym of "Drug", but we have 2 attributes under the Medication table with synonyms of "Drug", namely Med_Name and Med_Code. Since the NLQ has no further tokens to decide which attribute the user is referring to, we will output both of them.

For complex and nested queries like this example, the mapping and translation algorithm can be applied recursively.

Following the same steps of previous examples, we reach to the following acquired information:

- Table 1 = Medication
- Table 2 = Patient
- Attribute 1 = Med_Name
- Attribute 2 = Med_Code
- Attribute 3 = P_ID
- Attribute 4 = P_Name

And SQL query:

SELECT   Medication.Med_Name,   Medication.Med_Code   FROM Medication INNER JOIN Patient ON Patient.P_ID = Medication.P_ID WHERE Patient.P_Name = "Ahmed";

Figure 14 summarizes the steps taken to solve Example 4.

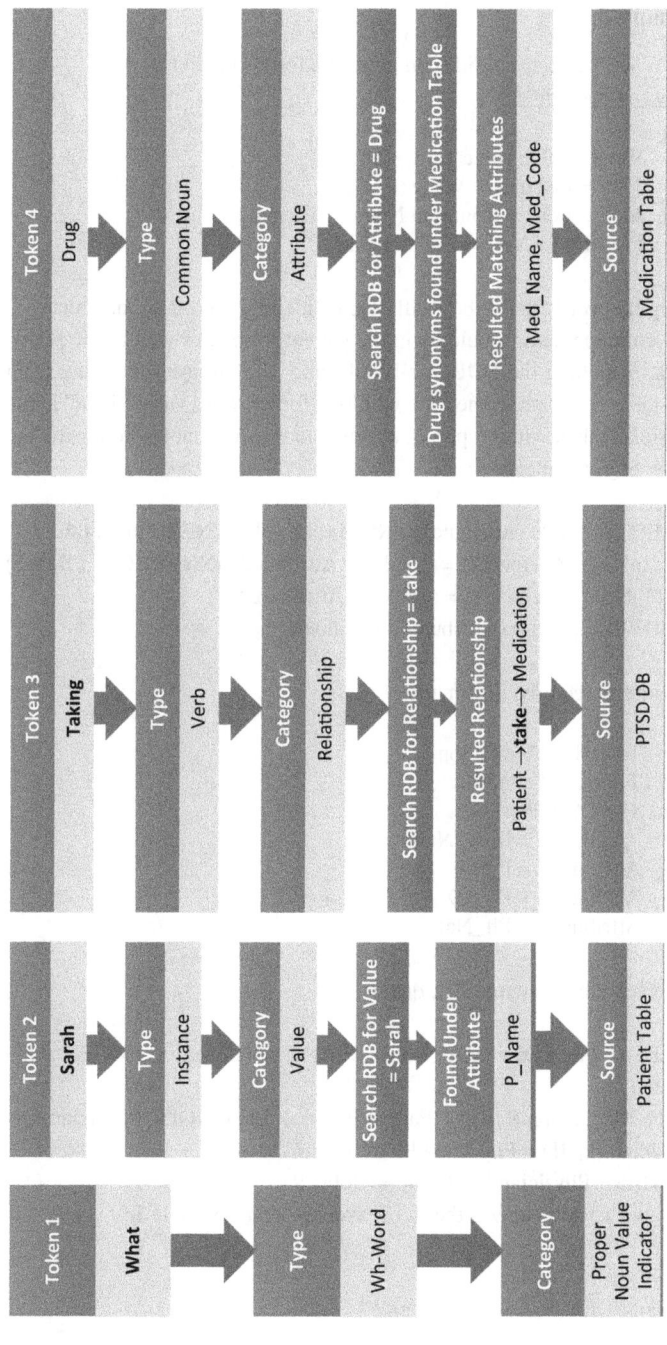

FIGURE 14 Example 4 Tokens Breakdown Analysis.

Example 5:

Q5: What medications did John prescribe for his patients?
  Tokens Breakdown:

- What =Value Indicator
- John = Instance = Value
- Medications = Common Noun = Attribute
- Prescribe = Verb = Relationship

Using the word "prescribe" will help us identify who is John, which is a very common name, could be in the patient's table, as well as the physicians. Searching the RDB, we'll only find 1 relationship pointing from physician to patient. Hence, we will look for the word value "John" in the Physician table, with proper join clauses to the three tables, following the below SQL template:

SELECT (Table1).(Attribute1) FROM ((Table1) INNER JOIN (Table2) ON (Table1).(Attribute2) = (Table2).(Attribute2) INNER JOIN (Table3) ON (Table3).(Attribute3) = (Table3).(Attribute3)
    WHERE (Table3).(Attribute4) = (Value);

With the acquired information:

- Table 1 = Medication
- Table 2 = Patient
- Table 6 = Physician
- Attribute 1 = Med_Name
- Attribute 2 = P_ID
- Attribute 3 = Ph_ID
- Attribute 4 = Ph_Name

We reach the following SQL query:

SELECT Medication.Med_Name FROM (Medication INNER JOIN Patient
    ON Medication.P_ID = Patient.P_ID) INNER JOIN Physician ON Physician.Ph_ID = Physician.Ph_ID)
    WHERE Physician.Ph_Name = "John";
    Figure 15 summarizes the steps taken to solve Example 5.

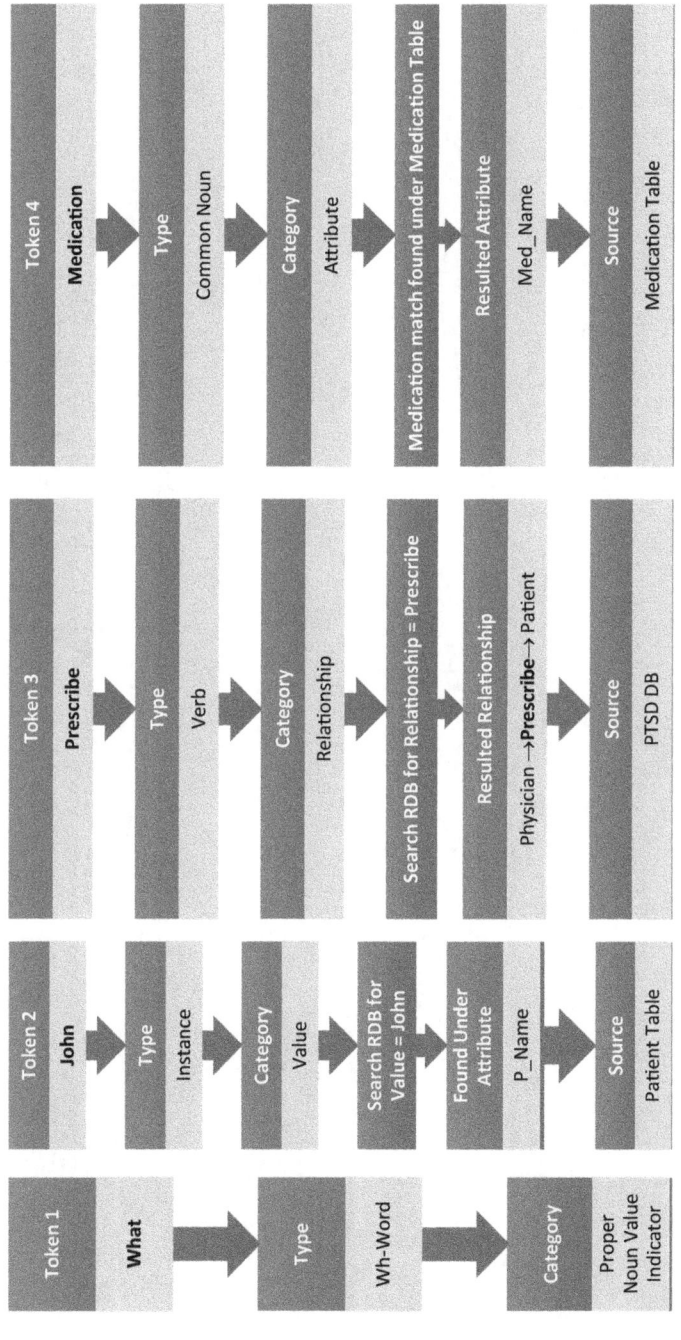

FIGURE 15   Example 5 Tokens Breakdown Analysis.

Implementation Testing and Performance Measurements

6

IMPLEMENTATION ENVIRONMENT AND SYSTEM DESCRIPTION

The machine used for this experiment is a MacBook Pro. It was used to run this experiment with macOS Mojave, version 10.14.2 (18C54). The processor speed is 2.9 GHz, Intel Core i7 (SATA Physical Interconnect), and 64bit architecture. The memory is 8 GB of RAM (distributed among two memory slots, each of which accepts a 1600 MHz memory speed and Double Data Rate 3 (DDR3) type of memory module), and 750 GB of disk space. The used MacBook has 1 Processor and 2 Cores, with 256 KB per core.

For the implementation coding and execution, Python 3.7 [164] was chosen as the programming language due to its clear syntax and popular NLP libraries for RDB processing tasks. The Integrated Development Environment (IDE) PyCharm C, Xcode and XQuartz were used to develop and compile the source codes as they have a Python unit-testing framework that allows for unit-testing automation in consistence with the Python Software Foundation [130]. The system's required dependencies include essential tools and supportive tools. All of the tools are downloaded and

DOI: 10.1201/b23367-6

installed locally on the experiment machine. The essential tools are declared in Figure 16, including:

- Python 3.7 [164]: A concise and lightweight programming language that is compatible with most OS platforms.
- MySQL Community Server 8.0.18 [165]: MySQL RDB backend server.
- MySQL RDB [166]: The RDB tool used to store and query data.
- NLTK [40]: Provides Python-compatible libraries for NLQ lemmatizing, tokenizing, tagging, parsing, classifying and semantic reasoning. It also supports interfaces to over 50 lexical resources in addition to WordNet corpora.
- TextBlob [167]: A Python library to process NLP tasks such as POS tagging, classification, noun phrase extraction and sentiment analysis.
- Stanford CoreNLP 3.9.2 [21]: Provides a set of integrated NLP tools to apply linguistic analysis on any incoming NLQ via a Python-compatible API. It offers sentences' structure sentiment analysis and syntactic and grammatical dependencies analysis. In addition,

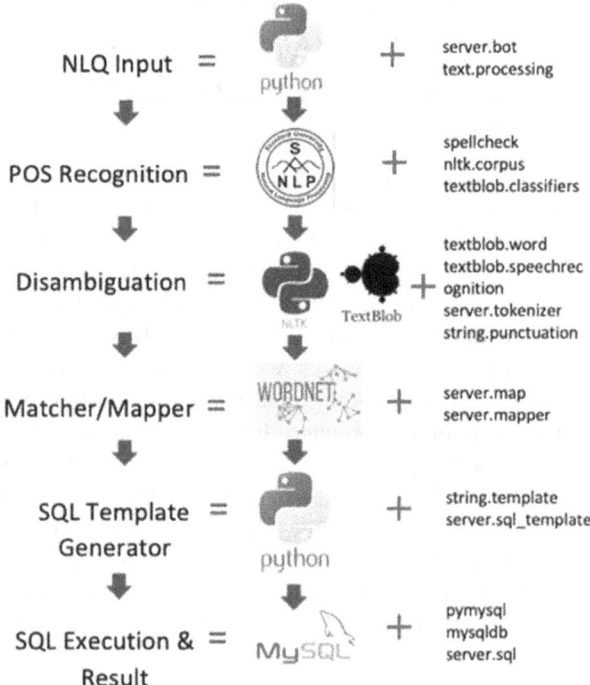

FIGURE 16   Framework structure, tools and libraries.

Stanford CoreNLP provides a stemmer, POS tagger, dates and times normalizer, NER, annotator, parser and bootstrapped pattern learning. Also, it offers the open information extraction tools such as extracting relationships between NLQ tokens.

- WordNet [19]: A large English lexical DB that includes nouns, verbs, adjectives, etc., in addition to the "synsets" library, which is a grouped set of cognitive synonyms.

The system's supportive tools include:

- IDE PyCharm C [168]: The Python IDE for code development and unit testing.
- XQuartz 2.7.11 [169]: A development environment designed for Apple OS X with supportive libraries and applications.
- Xcode 11 [170]: An application development tool for Apple OSX, used in this implementation to check codes' syntactic rightness.
- MySQL Workbench [171]: An SQL development and administration tool used mainly for visual modeling.

DATABASE

The current implementation uses MySQL DBMS as a backend environment. The implementation testing uses two RDBs, Zomato RDB [172] for algorithm testing, and the WikiSQL RDB [173] for algorithm validation. The testing process using a small RDB confirms the framework's functionality, while the framework validation process evaluates the framework's accuracy, efficiency and productivity.

Results from both Zomato (small RDB) and WikiSQL (large RDB) will be compared based on the RDB size. Table 11 compares between the two RDBs in terms of their number of instances or records, the number of tables and the public data source where they were published.

Zomato RDB [172], published in 2008, is a small RDB with a size of 2.5MB having 9,552 NLQ and SQL pairs stored in three comma-separated value (csv) file tables. Zomato RDB is about a restaurant search engine supplied by the public data platform "Kaggle". Zomato RDB has the schema demonstrated in Figure 17.

The WikiSQL_DEV RDB [174], published in 2017, was chosen because of its large RDB. It has 200.5 MB of 80,654 manually annotated RDB of NLQ and SQL pairs in 24,241 tables from Wikipedia. This RDB is used for developing NLIs for RDBs. Moreover, WikiSQL is considered the largest web-based

TABLE 11 Two RDBs comparison

	ZOMATO	WIKISQL
Size	2.5 MB	200.5 MB
NLQ/SQL Instances	9,552	80,654
Tables	3	24,241
Data Source	Kaggle	Wikipedia

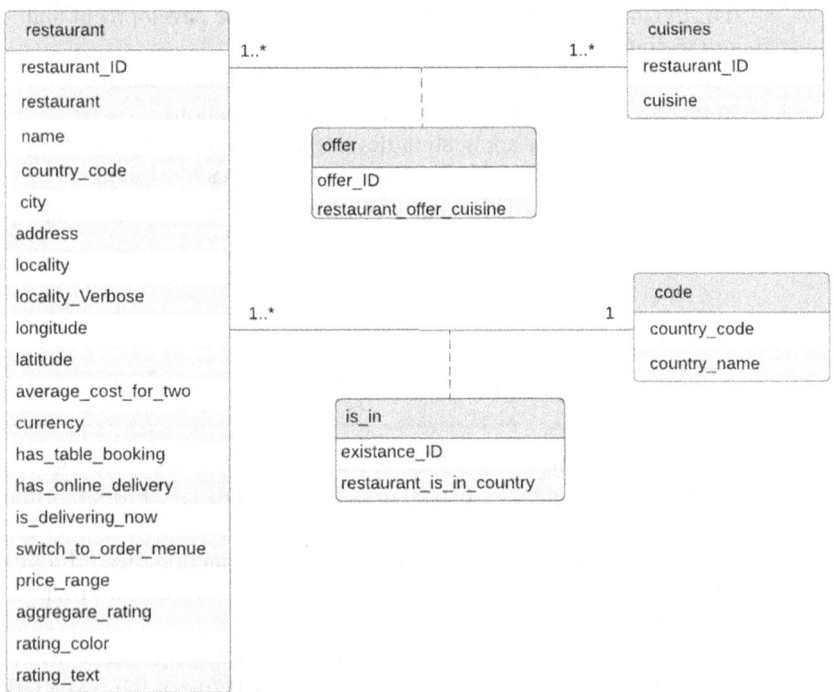

FIGURE 17 Zomato RDB schema.

realistic hand-annotated semantic parsing RDB [175]. This is because of the RDB's large magnitude and variety of logical form examples, tables, columns, lengths and types of questions, and the length of queries. Hence, it is the ideal RDB to generalize the implemented mapping algorithm to new and diverse queries and table schemata. Some examples of WikiSQL_DEV NLQ/SQL pairs are in Table 12.

TABLE 12   Examples of WikiSQL_DEV NLQ/SQL pairs

#	NLQ	TABLES	SQL
1	How many capital cities does Australia have?	"Country(exonym)", "Capital(exonym)", "Country(endonym)", "Capital(endonym)", "Official or native language(s) (alphabet/ script)"	SELECT COUNT (Capital (endonym)) FROM 1-1008653-1 WHERE Country(endonym)= Australia
2	What are the races that Johnny Rutherford has won?	"Rd", "Name", "Pole Position", "Fastest Lap", "Winning driver", "Winning team", "Report"	SELECT (Name) FROM 1-10706879-3 WHERE Winning driver=Johnny Rutherford
3	What is the number of the player who went to Southern University?	"Player", "No. (s)", "Height in Ft.", "Position", "Years for Rockets", "School/ Club Team/Country"	SELECT(No. (s)) FROM 1-11734041-9 WHERE School/Club Team/ Country=Southern University
4	What is the toll for heavy vehicles with 3/4 axles at Verkeerdevlei toll plaza?	"Name", "Location", "Light vehicle", "Heavy vehicle (2 axles)", "Heavy vehicle (3/4 axles)", "Heavy vehicle (5+ axles)"	SELECT (Heavy vehicle (3/4 axles)) FROM 1-1211545-2 WHERE Name=Verkeerdevlei Toll Plaza
5	How many millions of U.S. viewers watched the episode "Buzzkill"?	"No. in series", "No. in season", "Title", "Directed by", "Written by", "Original air date", "U.S. viewers(millions)"	SELECT COUNT (U.S. viewers (millions)) FROM 1-12570759-2 WHERE Title="Buzzkill"

IMPLEMENTATION TESTING AND VALIDATION

Testing the proposed mapping algorithm happens by running a randomized shuffling of the NLQ/SQL pairs from the Zomato RDB. This step uses four library functions namely, "random.shuffle", "collections.defaultdict", "tqdm" and "sql_parse.get_incorrect_sqls". First, the underlying NLP tools and the Matcher/Mapper module are tested by feeding the system the NLQ lemmatized tokens. Then, the tokens go through the Matcher/Mapper module to match the tokens with their synonyms built into the NLQ MetaTable. After that, tokens and their synonyms will be mapped to their adjacent RDB values, attributes, tables or relationships, each based on their syntactic role. To test the SQL template generator module, a set of RDB lexica will be passed to this module and the generated SQL will be examined for correctness, accuracy and other performance metrics discussed in the next section.

PERFORMANCE EVALUATION MEASUREMENTS

The purpose of the proposed algorithm is generating SQLs from NLQs automatically. It is important to obtain a reliable estimate of performance for this language translation algorithm. However, the algorithm's accuracy performance may rely on other factors besides the learning algorithm itself. Such factors might include class distribution, effect (cost) of misclassification and the size of training and test sets. Therefore, to validate the algorithm's performance and efficiency, more detailed accuracy measures are used to test the generated SQLs accuracy, precision and recall using:

- False Positive Ratio (FPR = $C/(C+D)$): the incorrectly classified queries as positives, but they are actually negatives.
- True Negative Ratio (TNR = $D/(C+D)$): the correctly classified queries as negatives.
- False Negative Ratio (FNR = $B/(A+B)$): the incorrectly classified queries as negatives, but they are actually positives.
- True Positive Ratio (TPR = $A/(A+B)$): the correctly classified queries as positives.

Where A = True Positive, B = True Negative, C = False Positive and D = False Negative.

The classification process here compares the generated SQL by the current framework against the designated SQL that is originally present in the testing RDB.

The recall performance measure represents the proportion of positive case (correct) queries which are correctly generated, Recall(R) = $A/(A+B)$. It also measures the presentation ratio of all relevant words by the system. The words here represent the derived lexica that are correctly identified from the RDB and lead to correct SQL generation. In this case, Recall = number of relevant words (lexica) retrieved from RDB/number of relevant words (lexica) not retrieved.

Precision is the proportion of the generated positive case (correct) queries which are correctly generated and considered as correct SQL constructions, Precision (P) = A/(A+C). It also measures how efficient the system is in retrieving only relevant words (lexica). In this case, precision is a measure of the ability of a system to retrieve and present only relevant lexica. Precision = number of relevant lexica retrieved/total number of lexica retrieved.

Moreover, results' correctness or accuracy is the proportion of total number of positive (correct) SQL generations which were correctly generated. Accuracy = (A+D)/(A+B+C+D). Unordered sets of retrieved queries can be evaluated by Precision and Recall. For ranked sets, after each query retrieval, precision should be plotted against recall.

The Receiver Operating Characteristics (ROC) curves [176] will also be used. ROC is a machine learning graphical plot of the TPR (a.k.a. sensitivity) against the FPR (a.k.a. 1-specificity). It makes a comparison between the two translation experiments in the current work, the experiment using the Zomato RDB and the second experiment using the WikiSQL RDB. This classification test takes into consideration the generated SQL by the current framework and the SQL already present in the testing RDB. The ROC curves show where the two experiment sets would possibly connect. This is because every TPR or FPR prediction instance is a single point on the ROC space. The bigger the area under the ROC curve the bigger the benefit of using the associated test. In other words, predictors' curves that are closer to the top-left corner provide better accuracy performance. Depending on the matching accuracy between the output SQL and the original NLQ input, the proposed algorithm is evaluated and documented.

In the first experiment with the Zomato RDB, 20 iterations (epochs) are executed on the system where the input NLQs are executed and their equivalent SQLs are generated as output. Then, the implementation resulted with the following performance metrics declared in Table 13 and Figure 18.

TABLE 13   First experiment confusion matrix with Zomato RDB

F-MEASURE: 94.5	ACTUAL POSITIVE	ACTUAL NEGATIVE
Predicted Positive	43 TPR (A)	3 FPR (C) Type I error
Predicted Negative	2 FNR (B) Type II error	49 TNR (D)
Accuracy: 94.85%	Recall: 0.96	Precision: 0.93

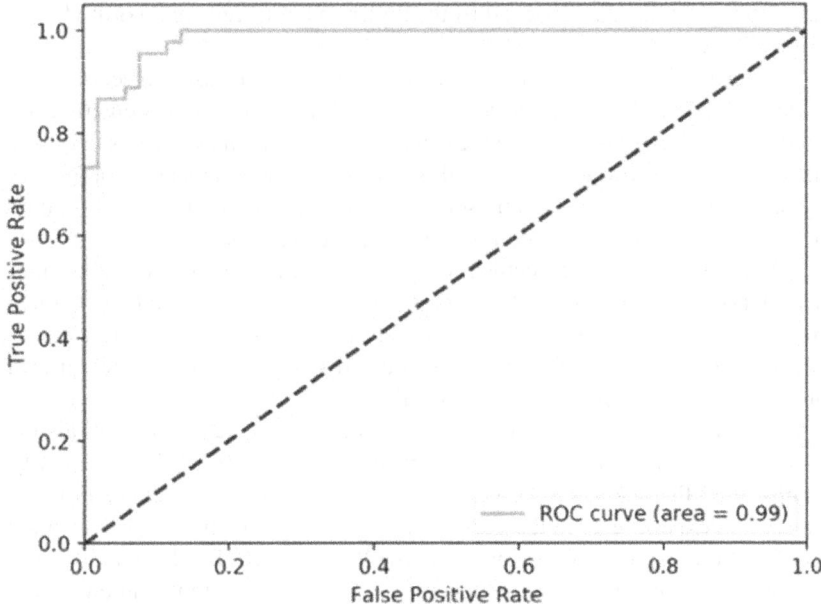

FIGURE 18   ROC curve for the first experiment with Zomato RDB.

For the second experiment with the WikiSQL RDB, the implementation resulted with the following performance metrics declared in Table 14 and Figure 19.

Compared with other similar research works on WikiSQL, the proposed work still achieves the highest accuracy measure as illustrated in Table 15 and Figure 20.

While the aforementioned performance measurements are sufficient to answer the current research question, the average time translating each query remains 1.5 minutes. This could be mainly due to the humble computer system

TABLE 14   Second experiment confusion matrix with WikiSQL RDB

F-MEASURE: 92	ACTUAL POSITIVE	ACTUAL NEGATIVE
Predicted Positive	42 TPR (A)	4 FPR (C) Type I Error
Predicted Negative	3 FNR (B) Type II Error	48 TNR (D)
Accuracy: 92.78%	Recall: 0.93	Precision: 0.91

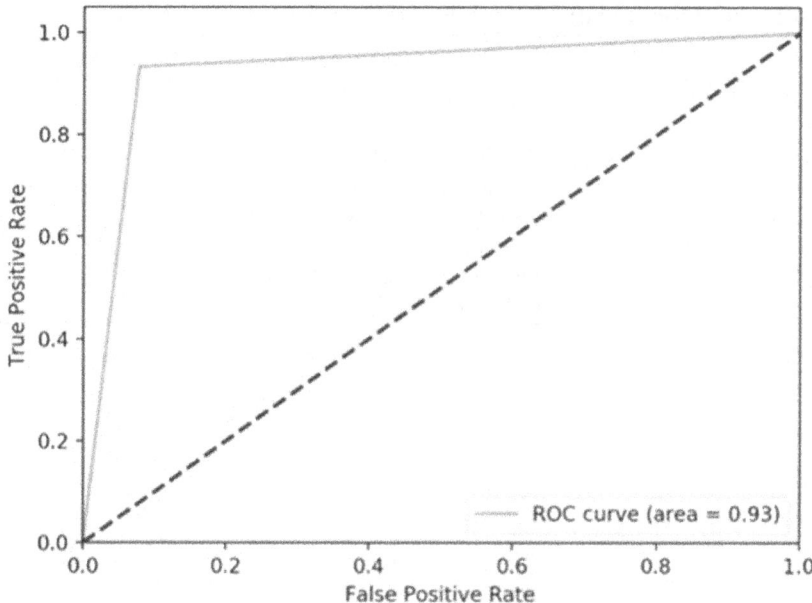

FIGURE 19   ROC curve for the second experiment with WikiSQL RDB.

TABLE 15   NLQ to SQL translation work on WikiSQL RDB

#	SOURCE	APPROACH	ACCURACY (%)
1	Proposed Algorithm	Computational Linguistics.	92.78
2	SEQ2SQL [177]	MLA through reinforcement Learning.	59.4
3	TypeSQL [174]	NLQ's token type recognition and 2 bi-directional LSTM.	82.6
4	SQLOVA [178]	Table- and context-aware NLQ word contextualization and representations.	89.6
5	X-SQL [179]	Reinforce schema representation with context.	91.8
6	WHERE clause variants [180]	Attentional Recurrent Neural Network (RNN).	88.6
7	DialSQL [181]	A dialogue-based framework that boosts the performance of existing algorithms via user interaction.	69

(Continued)

TABLE 15   (Continued) NLQ to SQL translation work on WikiSQL RDB

#	SOURCE	APPROACH	ACCURACY (%)
8	SQLNet [182]	A dependency graph, a sequence-to-set model and the column attention mechanism.	68.3
9	Question Patterns [183]	Question-pattern models containing dependency graphs.	Unmeasured
10	ValueNet [184]	A neural model based on an encoder-decoder architecture to synthesize the SQL query.	67

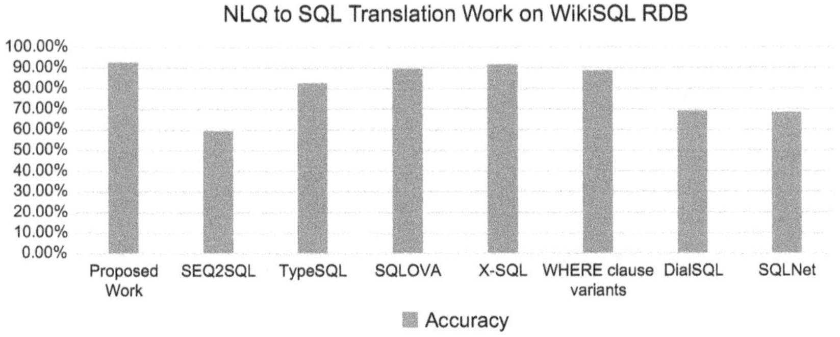

FIGURE 20   WikiSQL works accuracy comparison.

used to run the testing and validation processes. However, this processing time could be enhanced when analyzing the exact reasons of delay using further performance analysis. For example, each server executing the NLQ into SQL translation requests could be examined using the following specific performance metrics:

- Residence time, $RT = W/C$: the system's resource usage. This represents the average time queries spend in the server (actual service time + waiting time).
- Utilization, $U = B/T$: the average percentage of server's busy time.
- Throughput, $X = C/T$: the average percentage at which the server completes queries' translation requests.
- Queue length, $N = W/T$: the average number of queries at the server, whether executing the translations or waiting for service.
- Mean service time, $S = B/C$: the average time the server is busy with queries' translation processes.

- Area under graph, W = : the total time the server used to translate all queries.
  Where T = Total Period, C = Completed and B = Busy Time.

After running the analysis procedure, and as per Figure 21, it turns out that the process phase that took the longest time is the matching and mapping phase. This is to be expected since it does most of the tasks executed by the translation algorithm. A surprising discovery is the amount of time spent on the query execution and results retrieval from the MySQL RDB as follows:

- NL Interface = 0.04 secs
- POS Recognition = 0.25 secs
- Disambiguation = 0.06 secs
- Matcher/Mapper = 0.63 secs
- SQL Template Generator = 0.07 secs
- SQL Execution and Results = 0.45 secs

This time consumption breakdown represents the average time taken by each module to execute a translation task. They were computed after running a group of translation tasks and calculating the average time consumed by them combined.

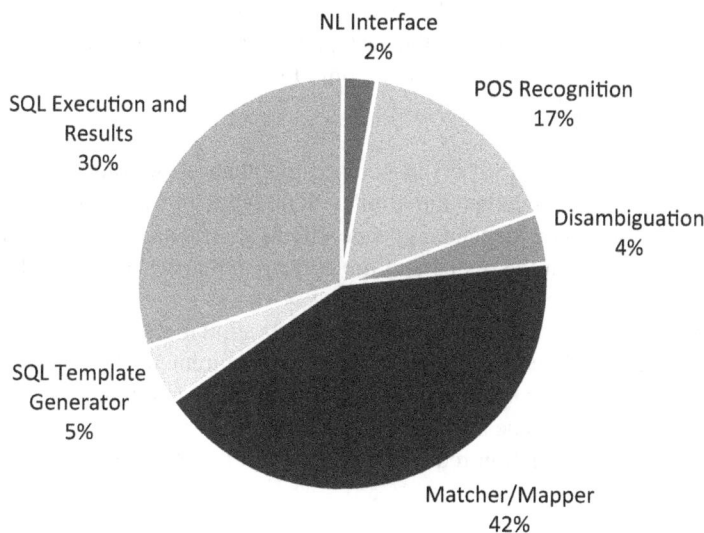

FIGURE 21   Translation average time distribution of 1.5 minutes.

When an SQL is passed to the DBMS engine for query processing, it is compiled by the MySQL Community Server compiler for result retrieval. This process is faster when dealing with small RDBs, but with large RDBs, the queries take much longer to execute. This is mainly because the MySQL RDB is not designed for big data application requirements.

The additional translation time consumed is proportional to the size of the RDB since the matching and mapping process will have to examine the whole RDB for potential matches. For example, with a 2MB RDB, the translation could take 5 seconds to complete, while it could take up to 1.5 minutes for a translation process to complete with an RDB of size 1GB.

What also affects the time consumption on the SQL execution and results phase is not having a query optimizer module. Since the same SQL query can be written in several ways, query optimization chooses the best way a query could be syntactically expressed. Not well-formed or optimized queries take longer to execute, affecting the overall process performance.

Finally, MySQL Community Server does not use proper temporary caching. Hence, frequently accessed data are not temporarily stored in the cache to insure faster future accesses. Consequently, the MySQL Server needs to establish a new service request whenever a new SQL query arrives for execution.

Generally, the server processing time consumed depends on:

- The number of queries.
- The amount of service each query needs.
- The time required for the server to process individual queries.
- The policy used to select the next query from the queue, the Queueing Model [185] (e.g., the First-Come-First-Served [186] or Priority Scheduling [187]).

The Queueing Model [188] could be used to enhance the translation performance by grouping together the similar SQL types in the queue. However, queueing models have dependency side effects considering the relationships between the SQLs in the queue and the corresponding service times for each SQL execution process.

Those effects could be mitigated by using similar calculations based on predicted workload intensity [189] and service requirements [190]. The workload intensity [189] is a measure of the number of translation requests made in a given time interval. The service requirements [190] represent the amount of time each query translation request requires from the server in the processing system.

If we assume that the system is fast enough to handle the arriving translation requests, the queries' translation completion rate (throughput) would equal the arrival rate. In this ideal case, the implementation environment would be called "jobs-flow balance" [191] where each query translation duration equals zero minutes instead of 1.5 minutes, as is the case in the current implementation environment.

Implementation Results Discussion

7

In terms of precision, Zomato RDB scored 93% while WikiSQL scored 91%. Those are the proportion of the correctly generated queries. It also measures the algorithm's efficiency in the identification and retrieval of matching RDB lexica since the retrieval of the wrong lexica would cause lower accuracy measures due to wrong SQL generations.

From the aforementioned precision and recall measures, the F-measure can be derived where F-measure = 2PR/ (P + R). The F-measure is the average performance measure of the matching RDB lexica retrieved as a result of an accurate matching and mapping process during the NLQ into SQL translation. Hence, the F-measure basically measures the accuracy of the data retrieved from an RDB as a result of applying an algorithm. The implementation execution using Zomato RDB had an F-measure of 94.5%, while WikiSQL had a 92%. In Table 16, a comparison between the two RDB experiments' performance measures is summarized in a confusion matrix. The numbers in the table represent averages over all runs of both experiments considering all queries in a run.

Figure 22 illustrates and summarizes all aforementioned performance metrics measures. With regard to the peaks of accuracy, recall, precision and F-measure bars in Figure 22, and in addition to the error rates (FPR and FNR) comparison, it can be concluded that the proposed algorithm is functioning properly and as needed. Hence, the mapping algorithm does indeed select the correct RDB elements successfully and map them with the correct SQL clauses using the novel mapping mechanism. This mapping is based on linguistics studies of the sentence structure by breaking down the sentence into the words level and study the words' inter-/intra-relationships.

Moreover, the area under the ROC curves, called AUC and shown in Figures 18 and 19, for Zomato RDB is almost 100% while WikiSQL AUC area is 93%. We conclude from this AUC comparison that the proposed algorithm shows accurate results for smaller RDBs, but not as much accuracy for RDBs that cover big data.

TABLE 16   Confusion matrix comparison between the two RDB experiments

MEASURE	ZOMATO RDB	WIKISQL RDB
Accuracy	94.85	92.78
Recall	96	93
TPR	43	42
FPR	3	4
TNR	49	48
FNR	2	3
Precision	93	91
F-Measure	94.5	92

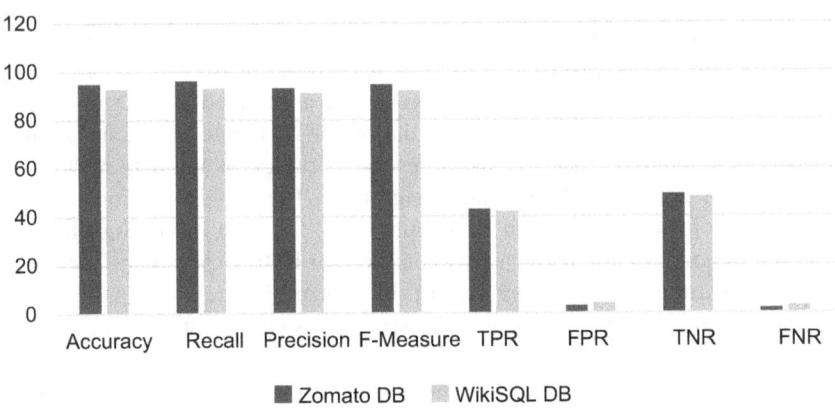

FIGURE 22   Comparison between the two RDB experiments.

IMPLEMENTATION LIMITATIONS

Mapping Limitations

The reason for the lack of accuracy in bigger RDBs, according to the proposed algorithm's experiments, is the system's confusion between the actual

RDB elements' names in the RDB MetaTable and the synonyms table as a whole. Thus, if a field is actually named "Birth_Day", and another field is named "BD" but has a synonym of "Birthday", the system will give priority to the field named "Birth_Day", which is the main source of confusion. However, the adoption of the synonym's table in NLP is quite immature and could be improved using appropriate machine learning techniques. Such techniques include classifying the synonyms and recognizing the actual column names.

Another cause of inaccuracy in the proposed framework is the mapping table. When NER or data profiling is used to import RDB's unique values and fields and tables' names, the algorithm will be obstructed from correctly mapping an NLQ token to an RDB value. This occurs when the NLQ token is not a unique value and therefore not included in the mapping table. Though the algorithm is supposed to search for the mentioned NLQ value in the whole RDB, it still starts with the RDB's unique values table (the mapping table) to minimize the searching time. This precedence prioritizes the RDB's unique values list, stored in the mapping table, over the entire RDB elements which increases the chances that the included unique values are mistakenly selected as a matching lexicon. Yet the value retrieval accuracy would still not be guaranteed since this depends greatly on how clearly the data clerk have entered the data and whether it had adequate synonyms attached to it.

SQL Generation Limitations

Other minor SQL generation errors, especially in the WikiSQL RDB, were due to the system's inability to grasp a thorough semantic understanding of the NLQ. An example is the system's inability to understand that the "king's speech" is the same as the "speech of the king". Other errors were related to SELECTing the wrong field due to NLQ's unresolved ambiguity.

In addition, the huge size of the WikiSQL RDB increases the likelihood of the presence of similarly named fields and tables which might even have identical labeling and synonyms. To mitigate this problem, RDBs must be preprocessed to have unique attribute and table names. Based on the proposed system and experiment results, the smaller the NLQ/SQL training patterns set the better results the system produces.

Larger sets have higher chances to cause rules conflictions or complications, while adding more rules would not always solve this conflict or increase results' accuracy. Translation failures could also be due to missing SQL clauses and not being retrieved because of ambiguous NLQs, or possibly due to mismatching in nested SQLs.

General Implementation Limitations

In the proposed system, the aim is building an automatic NLQ translation into an understandable language by a machine, that is SQL. However, NLIDBs require a significant amount of manual work for rule-based constraints, grammar specifications and RDB MetaTables annotation. In addition, RDB entities must be organized by properly naming tables and fields, converting all relationships' names into verbs, defining unique tables and identifying stored data types. With all said efforts, there is no need for a tailored data dictionary like most NLIDB question answering systems. This is because the designed MetaTables already have all knowledge-based facts needed for NLQs translation. On the other hand, the algorithm does require a significant amount of manual work in exchange, which can be worth it if the system will not need heavy future maintenance or customization along the way.

There are other general implementation limitations in this work, such as:

- The use of MetaTables as a Knowledge Base implies that the system is domain-dependent. This mandates MetaTables reconfiguration on any other RDB system, on which it is implemented, to be used by a DBA. The amount of manual work required is preprocessing the new RDB to have unique attribute and table names. Also, any acronyms or abbreviations must be excluded and changed to their proper namings as it is hard for the system to annotate acronyms with synonyms. The rest of the reconfiguration process is automatic in that the system has to process the RDB to be annotated and tagged with necessary metadata as illustrated in the RDB MetaTable.
- The Zomato RDB mapping table covers unique values of 5 fields and is about 12 KB, but with a larger DB, the mapping table becomes a limitation on the system resources (i.e., storage disks). As such, about 10% extra storage on large RDBS will be required to store all the metadata present in the RDB MetaTable.
- After executing SQL statements, the results returned are in a less human-understandable format; that is, columns and rows. Hence, the result should be manipulated to enable its presentation in an NLQ format. As an example, Yes/No questions, such as "Is there inpatients in the ICU Ward 5?", will return records if the answer is "Yes" and return no results if the actual result is "No"!
- The coding language used to implement such an algorithm must possess the capability to connect to an RDBMS and handle NLP tasks, i.e., Python or Java.

- Some queries are unanswerable and cannot be converted to SQL because they are too general, i.e., "Who is alive?" Another reason could be that the NLQ equivalent token in the RDB or its MetaTables cannot be mapped due to vague terms used in the NLQ sentence, i.e., "young" instead of "age between 4 to 7" or "heavy" instead of "weight more than 100 kilograms". The direct reason for unanswerable queries is the nonexistence of any matching records to the query conditions and constraints. In this case, the adoption of Fuzzy Logic solutions is necessary.

- Limitations regarding SQL language itself. For example, there is no loop operator or curser management capability in SQL. Also, expressing a universal quantifier in SQL requires a double-negated EXIST construct.

- Complicated or nested queries (i.e., aggregate functions) are not covered in the proposed algorithm as the model is tested on simple SELECT statements only. The current approach can be expanded to more complex queries. However, that might require an SQL construction module rather than an SQL template generator.

- The system has no embedded temporary memory (i.e., Cache or RAM) that stores past NLQs for accelerated frequent data retrieval. Temporary storage memory could also be used to store recently retrieved tables temporarily in case the user asks a follow-up question. It would be helpful if the temporary memory hosts related data until the user asks an NLQ of a different subject.

Conclusion and Future Work

8

CONCLUSION

It is almost impossible to find a technical application that does not require some sort of data storage and retrieval. The era of HCI, and the means of interaction, that is NLP, is in an ever-growing mode. In summary, the proposed algorithm aims at solving the language translation gap in the literature with the proposed mapping algorithm. The algorithm is designed to work on any RDB schema domain.

The mapping happens in accordance with a manually written rule-based mapping constraints algorithm. Two mappers are developed, one to map NLQ tokens into RDB elements, and another to map the identified RDB lexica into SQL clauses. The algorithm starts by analyzing the input NLQ by executing a series of NLP tasks. At each analysis stage, the data is further processed to lead to the formation and generation of an SQL. At the end, the SQL is executed, and the data are fetched from the RDB and displayed to the user.

The proposed algorithm covers many recent literature works' shortcomings using the following solutions:

i.   Limited HCI interaction with users to assure the most natural way of communication, that is direct questioning and answering. This way the user does not need to identify any NLQ tokens semantic roles.

ii. Does not need an annotated NLQ/SQL pairs corpus for training, making it domain-independent and adaptable on any environment.

iii. Uses simple algorithmic rules based on computational linguistics to fully understand the input NLQ for best translation accuracy.

iv.   Inventing NLQ and RDB MetaTables to increase mapping accuracy between NLQ tokens, RDB lexica, and then SQL clauses.

v.   Maps NLQ verbs with the RDB relationships as the simplest and most effective way of representing RDB elements relationships.

vi.   Overcomes any poor underlying linguistic tools' performance by using the supportive MetaTables and WordNet ontology.

vii. Supports NLQ's syntactic and semantic grammar analysis with computational linguistic algorithms (MetaTables constraints). Those constraints assist the NLQ tokens mapping to the RDB lexica, without using heavy-weighted and complicated techniques.

viii. Presents a significantly simpler algorithm as it relies on fewer, but more effective, linguistic tools and mapping rules without using intermediate representational layers.

To the best of our knowledge, this proposed research work for NLQ into SQL mapping and translation presents a novel mechanism. This work bridges the gap between RDBs and nontechnical DB administrators through a simple language translation algorithm using strong underlying NLP techniques. This work enables nontechnical users with no knowledge of RDB semantics to have the capability to retrieve information from employed RDBs.

The validation of the proposed research experiments and results has shown promising NLQ into SQL transformation and translation performance. As such, the smaller RDB performed a 95% accuracy, which is more than the larger RDB, which scored about 93%. This conclusion is in accordance with the applied performance metrics and measures such as accuracy, precision, recall and F-Measure.

However, larger RDBs in this experiment identified clear areas of improvement to enhance their language transformation accuracy to higher than a 93% accuracy. Another big area of improvement is further simplifying the algorithm coding and testing it on better implementation environment and technical resources. The aim is to minimise the translation time as it takes an average of 1.5 minutes to return a well-formed SQL, given an NLQ.

FUTURE WORK

Since the research around NLIDBs is only a few years old, there are so many future work opportunities to expand this work, including but not limited to:

- Adopting NoSQL queries.
- Scaling up to distributed storage RDBs.
- Employing community detection algorithms.
- Domain independent schema building.
- NLQ ambiguity and uncertainty resolution through fuzzy constraints.
- Dealing with vague and imprecise data through fuzzy RDBs that store fuzzy attribute values and fuzzy truth values.
- Attaching a well-designed user interface for NLQ input.
- Investigating neural network learning approaches for SQL ranking and classification based on a weighting scheme or an error/correctness rate.
- Processing NLQs with NLQ modifiers (i.e., almost, nearly, very).
- Processing NLQs expressed in NLQ time-stamped forms with prepositions (i.e., on, during, since).
- Outputting and transforming query translation and execution results into XML documents format. XML format is a standard scheme to store, interchange or exchange semi-structured to structured data.

Appendix 1

PSEUDOCODE 2 NLQ TOKENS LABELING

```
word = NLQ(tokens)
  for rslt = match_label[Table, Attribute, Value,
Relationship]
      if rslt[0] then
           label token as Table
      end if
      if rslt[1]
         if rslt[2] then
          label token as Value
          return(rslt[1], rslt[2])
      else
         label token as Attribute
      end if
      if rslt[3] then
         label token as Relationships
      end if
  end for
```

Appendix 1

Appendix 2

PSEUDOCODE 5 TOKENS' TYPE DEFINITION

```
for TextBlob(sentence[]) ← tokens
    if token type is noun_phrase then
        tag token as noun_phrase
        use as lexicon(table_name, attribute)
    if token type is string or number then
        tag token as Literal_Value
        use as lexicon(value)
    elif token type is proper_noun then
        tag token as proper_noun
        use as lexicon(value)
    elif token type is literal_value then
        tag token as literal_value
        use as lexicon(value)
    elif token type is verb then
        tag token as verb
        use as lexicon(relationship)
    elif token type is adverb then
        tag token as adverb
        use as lexicon(attribute)
    elif token type is adjective then
        tag token as adjective
        use as lexicon(attribute)
    elif token type is preposition then
        tag token as preposition
    elif token type is Wh_question then
        tag token as Wh_question
        use as lexicon reference
    elif token type is conjunction_phrase then
        tag token as conjunction_phrase
        use as lexicon condition
```

```
    elif token type is disjunction_phrase then
        tag token as disjunction_phrase
        use as lexicon condition
    elif token type is comparative_expression then
        tag token as comparative_expression
        use as lexicon condition
    else token type is operational_expression
        tag token as operational_expression
        use as lexicon condition
return sentence[tags]
```

Appendix 3

PSEUDOCODE 7 SYNONYMS TAGGING OF SQL
COMPARATIVE OPERATIONS KEYWORDS

```
keywords_synonyms()
    if keyword is average then
        add synonyms['average', 'avg']
    elif keyword is great then
        add synonyms['greater','gt','>','larger','more
than', 'is greater than']
    elif keyword is small then
        add synonyms['smaller','st','<','lesser
than','less than', 'is less than']
    elif keyword is greater_or_equal then
        add synonyms[greater or equal', 'gt or eq',
'>=', 'larger or equal', 'more than or equal']
    elif keyword is smaller_or_equal then
        add synonyms['smaller or equal', 'st or eq',
'<=', 'lesser than or equal', 'less than or equal']
    elif keyword is equal then
        add synonyms['equal', 'eq', '=', 'similar',
'same as', 'is']
    elif keyword is sum then
        add synonyms['what is the total', 'sum']
    elif keyword is max then
        add synonyms['what is the maximum', 'max',
'maximum']
    elif keyword is min then
        add synonyms['what is the minimum', 'min',
'minimum']
    elif keyword is count then
        add synonyms['how many', 'count']
    elif keyword is junction then
        add synonyms['and', 'addition', 'add',
'junction']
```

```
    elif keyword is disjunction then
        add synonyms['or', 'either', 'disjunction']
    elif keyword is between then
        add synonyms['among', 'between', 'range']
    elif keyword is order_by then
        add synonyms['order by', 'order', 'organise']
    elif keyword is asc then
        add synonyms['asc', 'ascending', 'small to
big', 'top to bottom']
    elif keyword is desc then
        add synonyms['desc', 'descending', 'big to
small', 'bottom up']
    elif keyword is group_by then
        add synonyms['group by', 'group']
    elif keyword is negation then
        add synonyms['negation', 'not', 'negative', 'is
not', 'are not', 'does not']
    elif keyword is like then
        add synonyms['what is the', 'like', 'similar
to', 'same as']
    else keyword is distinct
        add synonyms['distinct', 'unique']
    end if
return keywords_synonyms(tags)
```

Appendix 4

PSEUDOCODE 8 NLQ SPELLING CHECK FUNCTION

```
input = nlq(words)
output = correct_nlq(input)
while input ≠ Ø do
   if Spellcheck(nlq) = error then
      print ('Sorry, there is an error in your NLQ.',
'nlq')
      reset input = user_response(nlq(words))
      return input
      if ambiguitycheck(input) = true
         print out ('What did you mean by',
ambiguate(word), '?')
         classify user_response()
         if user_response = true then
            set input ← user_response(clarification)
         else
            set input = user_response(originalNLQ)
         end if
            end if
   else Spellcheck(nlq) ≠ error
   end if
end while
```

Appendix 5

**PSEUDOCODE 12 MATCHING NLQ TOKENS TO
EQUIVALENT RDB ELEMENTS**

```
for nlq(token) do
/* mapping tokens with their equivalent lexica or
their synonyms */
    if lexica[matching_lexicon(table, attribute,
value, relationship), synonym] ← token then
        find (matching_lexicon(table) -
    [HAS_ATTRIBUTE]-> matching_lexicon(attribute) -
    [HAS_VALUE]-> matching_lexicon(value)) -
    [HAS_RELATIONSHIP]->
    matching_lexicon(relationship)
        Compare token with matching_lexicon and
synonym
        spanTag matching_lexicon where
    matching_lexicon is similar to token and
    similarity > 0.75
        return matching_lexicon
    elif matching_lexicon > 1 then
        print ('which word did you mean to use?',
lexicon[0], 'or', lexicon[1])
        matching_lexicon ← user_response()
        return matching_lexicon
    else matching_lexicon[] ↮ token
        matching_lexicon(table) or
matching_lexicon(attribute) or
matching_lexicon(value) or
matching_lexicon(relationship) = False
        return error
    end if
/* find the corresponding RDB elements from the
identified ones */
```

```
    if   matching_lexicon(value) = True then
         matching_lexicon(attribute) ←
current_attribute
    elif matching_lexicon(attribute) = True then
         matching_lexicon(table) ← current_table
    else matching_lexicon(table) = True
         matching_lexicon(relationship) = current_
relationship
    end if
return matching_lexicon(table, attribute, value,
relationship)
replace token with matching_lexicon
end for
```

Appendix 6

PSEUDOCODE 13 BUILDING SQL MAIN CLAUSES

```
for sql_clauses(select, from, where) do

/* define where clause */
    if lexicon(attribute, value) = 1 then
        include select_clause(attribute, value)
    else lexicon(attribute, value) > 1 then
        include select_clause(attributes, values)
separated with ','
        return select_clause
    end if

/* define from clause */
    if lexicon(table) = 1 then
        include from_clause(table)
    else lexicon(table) > 1 then
        include from_clause(tables)
        join from_clause(tables)
        return from_clause
    end if

/* define where clause */
    where_clause ← condition_type[min, max, avg,
sum, count, distinct]
    if condition_type = True then
        include where_clause(conditions)
        add conditions with 'and'
        return where_clause
    end if

return select_clause + from_clause + where_clause
end for
```

113

Appendix 6

Appendix 7

Included SQL Query Types:

- Simple Queries:
  - SELECT 1 column in a table (or more) <u>without</u> conditions to present all data under selected column.
- Nested Queries (Subqueries):
  - SELECT 1 column in a table (or more) with WHERE condition\s.
- Cascaded Queries:
  - Join 2 or more columns FROM 2 or more tables in the SELECT/ FROM statement without conditions like:
    table-name1 JOIN table-name2 ON attribute1(PK of table1) = attribute2 (attribute in table2 and also FK of table1)
  - Join 2 or more columns FROM 2 or more tables in the SELECT/ FROM statement with WHERE conditions. The WHERE clause is a single condition or a joint of several conditions.
- Negation Queries:
  - Using the NOT Operator with SQL syntax to negate a WHERE condition.
- Simple WHERE Conditions:
  - 1 Simple operational condition (=,>, <, etc.)
  - 1 Aggregation condition (max, min, etc.)
  - 1 Negation Condition (NOT)
- Complex WHERE Conditions:
  - 2 or more operational conditions (=,>, <, etc.)
  - 2 or more Aggregation conditions (max, min, etc.) concatenated "=" with a value specified by the end user.
  - 2 or more Negation conditions (NOT AND)
  - Including subordinates and conjunctions
- Order/group by:
  - Asci. (Alphabetical, numeric).
  - Desc. (Alphabetical, numeric).

Appendix 7

Appendix 8

PSEUDOCODE 3 SQL TEMPLATE EXAMPLES

```
class Templates:
  /* zero attributes, one table */
  temp100 = Template('SELECT DISTINCT * FROM $table')
  /* zero attributes, one table, one attribute-value
pair */
  temp101 = Template('SELECT DISTINCT * FROM $table
WHERE $attribute='$value'')
  /* one attribute, one table */
  temp110 = Template('SELECT DISTINCT $attribute FROM
$table')
  /* one attribute, one table, two attribute-value
pairs (AND) */
  temp112 = Template('SELECT DISTINCT $attribute FROM
$table WHERE $attribute1='$value1' AND
$attribute2='$value2'')
  /* two attributes, one table */
  temp120 = Template('SELECT DISTINCT $attribute1,
$attribute2 FROM $table')
  /* zero attributes, two tables */
  temp200 = Template('SELECT DISTINCT * FROM $table1
NATURAL JOIN $table2')
  /* zero attributes, two tables, one attribute-value
pair */
  temp201 = Template('SELECT DISTINCT * FROM $table1
NATURAL JOIN $table2 WHERE $attribute='$value'')
  /* zero attributes, three tables, one attribute-
value pair (AND) */
  temp301 = Template('SELECT DISTINCT * FROM $table1
NATURAL JOIN $table2 NATURAL JOIN $table3 WHERE
$attribute='value'')
```

```
    /* zero attributes, three tables, two attribute-
value pairs (AND) */
    temp302 = Template('SELECT DISTINCT * FROM $table1
NATURAL JOIN $table2 NATURAL JOIN $table3 WHERE
$attribute1='$value1' AND $attribute2='$value2'')
    /* one attribute, three tables, two attribute-value
pairs (AND) */
    temp312 = Template('SELECT DISTINCT $attribute
FROM $table1 NATURAL JOIN $table2 NATURAL
JOIN $table3 WHERE $attribute1='$value1' AND
$attribute2='$value2'')
```

Appendix 9

TABLE 17 Literature works comparison

#	AREA	EXISTING SOLUTIONS	ADVANTAGE	DISADVANTAGE	HOW THESIS SYSTEM DIFFERS?
1	NLQ into SQL mapping Approaches	Authoring Interface Based Systems	• Uses semantic grammar specification, which is a language definition that provides accurate rules for linguistic expressions semantic parsing.	• Relies heavily on end-user input throughout multiple interface screens to modify the used keywords or phrases. • Requires extensive expertise time and efforts to identify and specify RDB elements and concepts.	• Only involves end users in the case of any NLQ words spelling mistakes or ambiguous phrases. • For linguistic expressions semantic parsing, NLP tools are used to lemmatize, tokenize, define and tag each NLQ token.
2		Enriching the NLQ/SQL Pairs via Inductive Logic Programming	• Widely used in MLA problems. • Provides logical knowledge and reasoning.	• Requires extensive manual rules defining and customizing in case of any DB change to maintain accuracy.	• The rule-based observational algorithm implemented is totally domain-independent and portable on any natural language translation framework. • Adds extra metadata to the NLQ/SQL pairs to easily find a semantic interpretation for NLQ's ambiguous phrases for accurate mapping.

3

Using MLA Algorithms

- NLQ/SQL pairs' corpora induces semantic grammar parsing to map NLQs into their SQLs.
- Used by training a Support Vector Machine (SVM) classifier, which is an efficient MLA for high dimensional datasets.

- Requires a huge domain specific NLQ/SQL translation pairs' corpora that is manually written.
- Data preparation is time consuming and a tedious task.
- Requires a domain expert to train and test the system.
- System is over-customized and unfunctional on any other domain.
- Assumes the user is familiar with the DB schema, data and contents.
- Relying heavily on MLAs are not effective in decreasing the translation error rates or increasing accuracy.
- SVM algorithm needs a lot of memory space.
- SVM is not scalable to larger DBs.

- Uses simple algorithmic rules and is domain independent.
- It does not assume prior knowledge of the adopted RDB schema or require any annotated corpora for training
- NLQ/SQL pairs are only used for algorithm testing and validation purposes.
- Focus is on understanding the NLQ to avoid potential future errors or jeopardize accuracy.

(Continued)

TABLE 17 (Continued) Literature works comparison

#	AREA	EXISTING SOLUTIONS	ADVANTAGE	DISADVANTAGE	HOW THESIS SYSTEM DIFFERS?
4		Restricted NLQ Input	• Uses a simple keyword-based search structure. • Uses a user-friendly form or template based or menu based NLI to facilitate the mapping process.	• Restricts the user to using certain domain-specific keywords. • Insignificant in terms of accuracy and recall. • Has portability problems even with advanced algorithms such as similarity-based Top-k algorithm.	• The current work facilitates the interaction between humans and computers without NLQ restrictions. • Has a limited interaction with the user to assure the most natural way of communication, direct questioning and answering, without needing the user to identify any NLQ tokens semantic roles. • Provides high accuracy and recall. • Compatible with any RDB domain and a translation environment.

5	Lambda Calculus	• Uses NLQs meaning representation for the mapping process. • A simple high-level language model of computation.	• Has some complicated language logic. • Too abstract in many cases. • Very slow in execution. • Hard to define rules with its logical expressions.	• Uses a compatible programming language, Python, that could be translated to any other language using grammatical parse trees and language compilers. • The current speed is an average of 1.5 mins per query.
6	Tree Kernels Models	• Applies kernel functions on NLQ/SQL pairs syntactic trees to learn its grammar. • Applies linear kernels on a "bag-of-words" to train the classifier to select the correct SQLs for a given NLQ.	• Requires a fully annotated NLQ/SQL pairs corpus. • Unable to recognise structural similarities and syntactic relations in an NLQ. • Has lower performance when scaled up to larger DBs.	• Does not require an NLQ/SQL pairs corpus to develop. • The employed NLP tools carefully understands the NLQ and recognizes its structural similarities and syntactic relations. • Has an insignificantly lower performance with larger RDBs as well but is still acceptable.

(Continued)

TABLE 17 (Continued) Literature works comparison

#	AREA	ADVANTAGE	DISADVANTAGE	HOW THESIS SYSTEM DIFFERS?
7	Unified Modeling Language (UML)	• Used to model the DB's static relationships and data models. • Refers to the DB's conceptual schema.	• Limited to a few class diagram concepts (e.g., classes, attributes, associations, aggregation and generalization). • The end user has to identify classes and their constituents. • UML models visualization requires compatible environments.	• MetaTables and mapping tables are used. They accommodate any type and kind of data. • Only NLQ input is required from the user. • Rule-based algorithm is compatible with all computational environments.
8	Weighted Links	• Uses the highest weight meaningful joins for mapping between NLQ tokens, RDB lexica and SQL clauses.	• Compromises accuracy with complexity. • Requires a huge annotated training dataset. • Computationally expensive.	• Accuracy and simplicity are both the main focus in the current work. • No training dataset is needed.

9	NLQ Tokens into RDB Lexica Mapping (NLQ Tokens Extraction)	Morphological and Word Group Analyzers	• Used for tokens extraction. • Analyses words' morphology.	• Requires a huge annotated training dataset. • Mapping accuracy is considerably low.	• The English word semantics dictionary (WordNet) is used to extract words' semantic information.
10		Pattern Matching	• Used to find keywords types. • Facilitates learning other domains' features.	• Requires a huge annotated training dataset. • Hard to analyse NLQ/SQL pairs mismatching causes.	• NLQ tokens extraction and their types identification happens through NLP computational linguistics processes, mainly the Lemmatizer and the tokenizer. • The assumption-based rules make it easy to find out causes of NLQ/SQL pairs mismatching.
11		Name Entity Recognizer (NER) Alone with Coltech-Parser in GATE	• Used to tokenize and extract NLQ's semantic information.	• Restricted to only recognize the NLQ tokens that already exist in the NER resource. • Integrating data from external resources is computationally expensive.	• While NER tagging is part of the underlying NLP tools, the main data source is acquired from the NLQ and RDB MetaTables.

(Continued)

TABLE 17 (Continued) Literature works comparison

#	AREA	EXISTING SOLUTIONS	ADVANTAGE	DISADVANTAGE	HOW THESIS SYSTEM DIFFERS?
				• Mapping accuracy is considerably low.	• MetaTables are used to check for tokens' existence as a first goal, then mapping them to their logical role as a relationship, table, attribute or value. • WordNet is used to support the MetaTables with words' synonyms, meanings and Lexical Analysis.
12		Java Annotation Patterns Engine (JAPE) Grammars	• Used for NLQ tokenization and NER tagging. • Generates a tag probability distribution. • Applies a rich feature representation.	• Less expressive than SQL-like languages. • It is memory extensive in that it creates a whole structured source tree for every DB element.	• No source trees are required except for the NLQ tokens' relations analysis step.

#	Method		
13	Porter Algorithm	• Used to extract tokens' stems. • Does not require knowledge structures' reprocessing.	• Only supports few languages. • Not a practical approach as it requires a huge memory to process. • Has a high false positives rate. • Hard to implement in other languages.	• The current work is language independent. • Does not mandate the availability of a huge memory, except for the storage of the RDB and the MetaTables.
14	Unification-Based Learning (UBL) Algorithm	• Extracts NLQ tokens using restricted lexical items and Combinatory Categorial Grammar (CCG) rules.	• Long processing time. • Complicated nature of stemmer. • Has a high error rate in recognizing NLQ noun phrases. • Difficulty in analyzing tokens relations.	• Average processing time of 1.5 mins per query. • The NLP tools easily identify and recognize tokens' semantic roles and their lexical relations.
15	Dependency Syntactic Parsing	• Used to extract tokens and their lexical relations. • Replaces parse trees with dependency structures. • Captures meaningful dependency relations directly.	• Potential data loss during parse tree generation and expansion. • Eventual error propagation while applying the greedy parsing. • Language dependent.	• NLTK parser parses the NLQ tokens according to the built-in semantic roles that are mapped to specific RDBs elements. • A parse tree is generated and a dictionary of table names, attributes and tokens are maintained, and NLQ's subjects, objects and verbs are identified.

(Continued)

TABLE 17 (Continued) Literature works comparison

#	AREA	EXISTING SOLUTIONS	ADVANTAGE	DISADVANTAGE	HOW THESIS SYSTEM DIFFERS?
16		Separate Value and Table Extractor Interfaces	• A compromising approach for not supporting the RDB schema elements' MetaTables and synonyms. • Ideal for complex NLQ/SQL pairs.	• Requires a big annotated training dataset. • Does not provide information on NLQ tokens' semantic relationships. • Requires a long time to process.	• Supports RDB schema elements' MetaTables and synonyms for tokens' semantic information. • No need for a rich annotated corpus of NLQ/SQL pairs for algorithm training. • Domain-independent and configurable on any working environment.
17	NLQ Tokens into RDB Lexica Mapping (RDB Lexica Mapping)	Spider System	• Uses a rich corpus created using complex and cross-domain semantic parsing and SQL patterns coverage. • An incremental approach, new experiences affect processing.	• Uses a huge human labeled NLQ/SQL corpus for training and testing. • Mapping accuracy is not significantly high.	• Focuses on simplicity and accuracy of the algorithm's mapping outcome with highest priority. • Uses NLQ MetaTable to map NLQ tokens into RDB lexica. • The implemented MetaTables fill up the low accuracy gap in language translation algorithms that do not use any sort of deep DB schema data dictionaries or just a limited Data Dictionary.

| 18 | WordNet alone | • Efficient at expanding NLQ predicate arguments to their meaning interpretations and synonyms.
• Handles complex NLQs without an ontological distinction. | • Generalizes the relation arguments and does not guarantee NLQ's lack of ambiguity and noise which significantly affects its meaning interpretation. | • Supportive techniques are employed in the current research work such as the disambiguation module.
• To avoid confusion around the RDB unique values, data profiling is performed on RDB's statistics to automatically compile the mapping table of unique values, PKs and FKs.
• Unique values, PKs and FKs are stored in a mapping table by specifying their hosting attributes and tables while a hashing function is used to access them instantly. |

(Continued)

TABLE 17 (Continued) Literature works comparison

#	AREA	EXISTING SOLUTIONS	ADVANTAGE	DISADVANTAGE	HOW THESIS SYSTEM DIFFERS?
19	NLQ Tokens into RDB Lexica Mapping (RDB Lexica Relationships)	Stanford Dependencies Parser	• Can parse any language in any free word order. • Displays all sentence structure and tokens dependencies. • Uses parsing trees to represent syntax and semantics.	• It is an outdated lexicalized parser, which leads to unnecessary errors. • Sentences must follow Chomsky Normal Form (CNF) style.	• NLQ can be in any form as long as it has correct spellings and no ambiguous tokens.
20		Dependency Syntactic Parsing	• Simple and expressive. • Displays each token in the NLQ in a high level.	• Does not show any semantic information. • Some parsing trees are erroneous in that they never lead to the targeted RDB elements. • Potential false early prune out.	• The simplest and most effective way of representing RDB elements relationships is by restricting the RDB schema relationships to be in the form of a verb for easy mapping between NLQ verbs and RDB relationships.
21		Dependency-Based Compositional Semantics (DCS) System Enriched with Prototype Triggers	• Parsing and learning is done using logical forms (trees). • Has rich a properties set of computations, statistics and linguistics.	• Requires manual annotation of logical forms and semantic parsing. • Complex implementation.	• Simple rule-based algorithm that uses the semantic role of a verb to link RDB lexica with each other. • No manual annotation is needed.

22	NLQ Tokens into RDB Lexica Mapping (NLP syntax and semantics)	Named Entity Tagger	• Recognizes a wide range of literal values of named or numerical entity sets.	• Does not remember previously tagged entity sets. • Supports limited languages. • Does not show dependencies between named entity sets.	• Previously tagged entity sets are saved (temporarily in case of limited storage) in the NLQ MetaTable. • Supports the NLQ's syntactic and semantic grammar analysis with computational linguistics algorithms in the form of RDB and NLQ MetaTables to assist tokens mapping into RDB lexica. • NLP syntactic and semantic tools show the source tables of each token, which explains tokens' relationships.
23		Dependency Parser	• Produces parallel syntactic dependency trees. • Constructs dependency trees directly without any parse trees conversions.	• Does not recognise complex language phenomena.	• Considers understanding the NLQ, by finding the combination of its tokens' meanings, is the most essential part in the mapping and translation process.

(Continued)

TABLE 17 (Continued) Literature works comparison

#	AREA EXISTING SOLUTIONS	ADVANTAGE	DISADVANTAGE	HOW THESIS SYSTEM DIFFERS?
				• Employs computational linguistic studies at the words processing level is employed. • The current research discovered common semantics between NLQ and SQL by analyzing the language syntax roles.
24	LIFER/LADDER Method	• Uses NLQ syntactic and semantic analysis alone. • Simple and easy to implement.	• Not sufficient and produces substantially low precision, FPR and TNR.	• This research framework overcomes any poor underlying linguistic tools' performance that are meant to analyse NLQ syntax and semantics by using the supportive MetaTables and WordNet ontology. Such NLP tools include named entity tagger, tokenizer, or dependency parser.

25	NLQ/SQL Syntax Trees Encoded Via Kernel Functions	• Learns multiple NLQ syntactic features. • Represents unrestricted features of domain-specific knowledge.	• Limited to available data resources. • Expensive development and high time consumption. • Does not show dependencies between named entity sets.	• RDB schema knowledge, the semantic data models in the form of MetaTables, and syntactic-based analysis knowledge are used to generate parse trees from the identified tokens to properly map NLQ tokens to the related RDB elements.
26	The Probabilistic Context Free Grammar (PCFG) Method	• Models NLQ features using production rules with estimated probabilities. • Uses overlapping and interdependent features to build its probability models.	• Proved to be challenging in terms of finding the right grammar for optimization. • Iterative production rules lead to inherited computational complexity. • Requires an annotated training and testing dataset.	• The current research discovered common semantics between NLQ and SQL by analyzing the language syntax roles. • Does not require annotated datasets.

(Continued)

TABLE 17 (Continued) Literature works comparison

#	AREA	EXISTING SOLUTIONS	ADVANTAGE	DISADVANTAGE	HOW THESIS SYSTEM DIFFERS?
27	RDB Lexica into SQL Clauses Mapping (SQL clauses mapping)	The Extended UML Class Diagrams Representations	• Extracts fuzzy tokens semantic roles in a form of a validation sub-graph or tree of the Self Organizing Maps (SOM) diagram representation that transforms class diagrams into SQL clauses using the fuzzy set theory. • More flexible than the MLA approaches. • Provides higher measures of recall.	• Has a high False Positive Ratio.	• Uses computational linguistics mapping constraints to transform lexica into SQL clauses and keywords. • Computational linguistics is used here in the form of linguistics-based mapping constraints using a manually written rule-based algorithm. • Those algorithms are mainly observational assumptions. • The MetaTable specifies RDB schema categories (value, relationship, attribute, etc.) to map the identified RDB lexica into SQL clauses and keywords. • Provides high measures or accuracy and recall.

| 28 | RDB Relationships and Linguistic Analysis | • Relationships are used to map the lexica into NLQs linguistic semantic roles' classes as a conceptual data model.
• The mapping results are derived from the connectivity matrix by searching for any existing logical path between the source (objects) and the target (attributes) to eventually map the logical path to an equivalent SQL. | • The user has to identify the source, its associations or relationships and the target in the fuzzy NLQ to connect them for UML class diagram extraction.
• Extraction is not thorough or exhaustive. | • Current work uses RDB relationships and NLP tools, which are more capable of "understanding" the NLQ statement before translating it to an SQL query.
• This method highly contributes to the increase in the translation accuracy.
• Regarding the linguistic inter-relationships within the RDB schema in the current work, not only WordNet is used, but also an NLTK and NLP tools. Besides, a manual rule-based algorithm is also used to define how NLQ linguistic roles match with the RDB elements, which explains the variance in translation accuracy and precision in comparison. |

(Continued)

TABLE 17 (Continued) Literature works comparison

#	AREA	EXISTING SOLUTIONS	ADVANTAGE	DISADVANTAGE	HOW THESIS SYSTEM DIFFERS?
					• Assures a seemingly natural interaction between the user and the computer. Hence, the user does not have to identify any semantic roles in their NLQ. The underlying NLP tools does this for them. • The relationships are identified by the NLQ verbs, so the user is communicating more information in their NLQ using the current research algorithm compared to the other literature works. Hence, it is considered more advanced and user friendly. • Not only objects and attributes are extracted from the NLQ, the proposed research work extracts much lower level

| 29 | RDB Lexica into SQL Clauses Mapping (Complexity vs Performance) | L2S System | • Compares all existing NLQ tokens with existing DB elements using NLP tools, tokens' semantic mapper and a graph-based matcher. | • A complicated system that consumes a lot of time to run through all DB elements for comparison with NLQ tokens.
• Computationally expensive. | linguistic and semantic roles such as gerunds and prepositions which help select the matching RDB lexica with higher accuracy and precision.
• Considered significantly simpler than most complex mapping approaches as it relies on fewer, but more effective, underlying linguistic tools and mapping rules. |
| 30 | | Bipartite Tree-Like Graph-Based Processing Model | • Employs sophisticated semantic and syntactic analysis of the input NLQ. | • A complicated system that requires a domain-specific background knowledge and a thorough training and testing dataset. | • The current work is the best in terms of performance, simplicity and adaptability to different framework environments and RDB domains.
• The use of MetaTables to define the lexica semantic roles and their adjacent SQL slots for better mapping accuracy. |

(Continued)

TABLE 17 (Continued) Literature works comparison

#	AREA	EXISTING SOLUTIONS	ADVANTAGE	DISADVANTAGE	HOW THESIS SYSTEM DIFFERS?
31		Ellipsis Method	• Deals with instances of ellipsis (less than a sentence). • Uses a computationally cheap and robust approach.	• Requires that NLQ be explained by the user. • A memory-based learning method. • Produces a few mismatched SQLs.	• Employs a lightweight approach for query translations with no need to storage spaces other than storing the MetaTables and the mapping tables.
32		The Highest Possibility Selection	• Automatically discards any features that are not necessary. • Memorizes and searches previous NLQ encounters to find relatable features.	• Relies heavily on labeled training and testing data, which is expensive and tedious to create. • Not generalizable across other domains.	• No need for labeled data for developing and executing. • Can be generalizable across domains.
33		Weighted Neural Networks and Stanford Dependencies Collapsed (SDC)	• Generates ordered and weighted SQLs schemata. • Uses linguistics in their algorithm.	• Computationally expensive. • Unscalable to bigger RDBs.	• The translation accuracy of this algorithm still falls behind the proposed algorithm because of the use of further semantic roles and linguistic categories (i.e., adjectives, pronouns etc.).

| 34 | Pattern Matching of SQL | • Uses NLQ's subject or object to search the DB for matching attributes with weighted projection-oriented stems and generate the SQL clauses accordingly.
• Used as a mapping algorithm.
• Easy to develop and execute.
• Handles mixed data types of NLQ tokens. | • Prioritizes SQLs based on probability of correctness instead of accuracy and precision.
• Does not apply any NLQ interpretation modules or parsing elaborations.
• Its translation accuracy and overall performance is highly jeopardized. | • Uses the verbs to find the attributes' and values' relationships instead of using a heavy weighted tool such as the weighted projection-oriented stems.
• Both implemented mappers have access to an embedded linguistic semantic-role frame schema (WordNet and Stanford CoreNLP Toolkit), data dictionary (MetaTables) and the RDB schema. Those resources are essential for accurate SQL query formation and generation. |

(Continued)

TABLE 17 (Continued) Literature works comparison

#	AREA	EXISTING SOLUTIONS	ADVANTAGE	DISADVANTAGE	HOW THESIS SYSTEM DIFFERS?
35	RDB Lexica into SQL Clauses Mapping (SQL Formation vs SQL Templates)	NLQ Conceptual Abstraction	• A concept-based query language to facilitate SQL formulation. • Scalable to large datasets.	• SQLs are constructed from scratch, which adds extra computational complexity to the language translation system. • Adds an additional unnecessary layer on top of the original system architecture.	• Simplifies SQL queries generation by using ready SQL templates. • SQL construction constraints are used in the mapping algorithm to guarantee accurate SQL template selection. • This approach is considered as a simple and accurate method of generating SQLs.
36		Semantic Grammar Analysis	• Used to store all grammatical words to be used for mapping NLQ's intermediate semantic representation into SQL clauses.	• Due to this system's complexity, this architecture can only translate simple NLQs. • Not flexible with nested or cascaded SQLs.	• Accommodates more complex SQL types such as nested or cascaded SQLs.

37

| Kernel Functions, SVM Classifier, and the Statistical and Shallow Charniak's Syntactic Parser | • Used to classify NLQ/SQL pairs as correct or incorrect.
• The mapping algorithm is at the syntactic level.
• Uses NLQ semantics to build syntactic trees to select SQLs according to their probability scores.
• A parser is applied to compute the number of shared high-level semantics and common syntactic substructures between 2 trees to produce the union of the shallow feature spaces. | • Achieves low recall of correctly retrieved SQL answers.
• Requires labeled training and testing datasets of NLQ/SQL pairs.
• Such exclusive domain-specific systems are highly expensive.
• Its performance is subjective to the accuracy and correctness of the training and testing datasets, which are manually written by a human domain expert. | • No need to develop a training and testing datasets of NLQ/SQL pairs for every new domain.
• High recall due to the use of an accurate mapping algorithm mapped to ready SQL templates. |

(Continued)

TABLE 17 (Continued) Literature works comparison

#	AREA	EXISTING SOLUTIONS	ADVANTAGE	DISADVANTAGE	HOW THESIS SYSTEM DIFFERS?
38		Heuristic Weighting Scheme	• NLQ/SQL pairs syntactic trees are used as an SQL compiler to derive NLQ parsing trees. • NLQ tokens' lexical dependencies, DB schema and some synonym relations are used to map DB lexica with the SQL clauses.	• There is no use of any NLQ annotated meaning resources or manual semantic interpretation and representation to fully understand the NLQ. • The SQL generator performance is considerably low. • SQL generator is built from scratch, which adds high complexity to the language translation algorithm.	• Uses simple algorithmic rules based on computational linguistics to fully understand the input NLQ to ensure highest translation accuracy. • RDB MetaTable is used for lexical relations disambiguation. • A mapping table is also used, which includes RDB lexica data types, PKs and FKs and names of entity sets (unique values), in addition to other rule-based mapping constraints. • Uses SQL templates and puts extra focus on passing accurate RDB lexia into SQL templates generator for better performance and correct output.

39	A Deep Sequence to Sequence Neural Network	• Generates an SQL from NLQ semantic parsing. • Uses reinforcement learning and rewards from in-the-loop query execution to learn an SQL generation policy.	• Incompatible with cross entropy loss optimization training tasks. • Requires manually annotated NLQ/SQL pairs for generating the SQL conditions. • Execution accuracy is as low as 59.4% and logical form accuracy is 48.3%. • Proved to be inefficient and unscalable on large RDBs.	• Uses manually written rule-based grammar for the mapping. • Produces high measures of accuracy and recall. • No need for training and testing labeled data.
40	MLA Sequence-To-Sequence-Style Model	• Employs a mapping algorithm without reinforcement learning. • Showed small improvements to generate SQL queries when order does not matter. • Solves the "order-matters" design problem.	• Has very low performance measures. • Had to use dependency graphs and the column attention mechanism for performance improvement. • The model has to be frequently and periodically retrained to reflect the latest dataset updates, which increases the system's maintenance costs and computational complexity.	• Translates NLQs into SQLs while maintaining high simplicity and performance. • The system does not need to be updated or maintained periodically.

(Continued)

TABLE 17 (Continued) Literature works comparison

#	AREA	EXISTING SOLUTIONS	ADVANTAGE	DISADVANTAGE	HOW THESIS SYSTEM DIFFERS?
41		A Deep-Learning-Based Model	• Predicts and generates the SQL directly for any given NLQ. • Uses attentive-copying mechanism, a recover technique and task-specific look-up tables, to edit the generated SQL. • Overcomes the shortcomings of sequence-to-sequence models. • Proved its flexibility and efficiency	• NLQ/SQL pairs were manually written for model training and testing. • The used RDB is specifically customized to the used framework and environment applied on. • Highly questionable in terms of generalizability, applicability and adaptability on other domains.	• Uses RDBs that are public source namely, Zomato and WikiSQL, only for algorithm validation and testing. • Does not need labeled data for developing. • The translator algorithm is domain-independent and configurable on any other environment.

| 42 | RDB Lexica into SQL Clauses Mapping (Intermediate Representation) | Regular Expressions (regexps) | • Represents NLP tokens phonology and morphology.
• Uses NLQ intermediate semantic representation layers to represent NLQ lexica as SQL clauses.
• Tokens representation happens by applying First Order Predicate Calculus Logic resembled by DB-Oriented Logical Form (DBLF) and Conceptual Logical Form (CLF) with some SQL operators and functions to build and generate SQLs. | • regexps collections in NLQ sentences are not clearly articulated in the literature.
• Proved to be not as effective as the NLP tools, MetaTables and mapping tables in terms of accuracy and precision. | • Uses NLQ MetaTables and RDB MetaTables to increase accuracy of mappings between NLQ tokens into RDB lexica and then into SQL clauses.
• Tries to save every possible information given by the NLQ so as each NLQ token is used and represented in the SQL clauses and expressions production.
• Uses multiple NLP tools, MetaTables and mapping tables for unique values to fully understand the NLQ and map its tokens to their corresponding RDB elements. |

(Continued)

TABLE 17 (Continued) Literature works comparison

#	AREA	EXISTING SOLUTIONS	ADVANTAGE	DISADVANTAGE	HOW THESIS SYSTEM DIFFERS?
43		The Similarity-Based Top-K Algorithm	• Processes NLQ tokens to map them to their internal conceptual representation layer. • Uses Entity-Attribute-Value (EAV) DB metadata and grammatical parse trees.	• Proved to be not effective because of the high complexity and time consumption approaches applied.	• No conceptual representation is needed for the mapping. • The identified attributes are automatically mapped into the SQL SELECT clause, while the tables are extracted from the SELECT clause to generate an SQL FROM clause, and the values are used as a conditional statement in the WHERE clause.
44		Lambda-Calculus	• Uses lambda-calculus to map tokens to their corresponding meaning representations.	• A supervision-extensive system.	• No meaning representation is needed for the mapping. • Does not require any human supervision to properly function.
45		An Intermediate Tree-Like Graph	• Transforms DB lexica into an intermediate tree-like graph. • Extracts the SQL from the maximum bipartite matching algorithm.	• Computationally expensive and processing is time consuming.	• No internal graphical representation is needed for the mapping. • Processing the mapping has an average speed of 1.5 mins.

Glossary

AI	Artificial Intelligence
API	Application Program Interface
CSV	Comma-Separated Values
DAC	Data Administration Commands
DB	DataBase
DBMS	Database Management System
DCL	Data Control Language
DDL	Data Definition Language
DML	Data Manipulation Language
DQL	Data Query Language
EAV	Entity-Attribute-Value
ERD	Entity-Relational Diagram
FKs	Foreign Keys
FNR	False Negative Ratio
FPR	False Positive Ratio
IDE	Integrated Development Environment
MLA	Machine Learning Algorithm
NER	Named Entity Recognition
NL	Natural Language
NLI	Natural Language Interface
NLIDB	Natural Language Interface for DataBase
NLP	Natural Language Processing
NLQ	Natural Language Question
NLTK	Natural Language Toolkit
NLTSQLC	NL into SQL Convertor
NoSQL	Not Only Structured Query Language
OLTP	Online Transactional Processing
PKs	Primary Keys
POS	Part of Speech
PTSD	Post-Traumatic Stress Disorder
QAS	Question Answering Systems
QL	Query Language
RA	Relational Algebra
RDB	Relational DataBase

RDBMS	Relational Database Management Systems
ROC	Receiver Operating Characteristics
SQL	Structured Query Language
TCL	Transaction Control Language
TNR	True Negative Ratio
TPR	True Positive Ratio

References

[1] Nihalani, N. (2010). An intelligent interface for relational databases. *Human–Computer Interaction*, *6*, 7.

[2] Natural language processing (NLP), Wikipedia. Last accessed March 29, 2020. https://en.wikipedia.org/wiki/Natural_language_processing

[3] Kaur, J., Chauhan, B., & Korepal, J. K. (2013). Implementation of query processor using automata and natural language processing. *International Journal of Scientific and Research Publications*, *3*(5), 1–5.

[4] Bronnenberg, W. J. H. J., Landsbergen, S., Scha, R., Schoenmakers, W., & van Utteren, E. (1978). PHLIQA-1, a question-answering system for data-base consultation in natural English. *Philips Technical Review*, *38*, 229–239.

[5] Aron, J. (2011). How innovative is Apple's new voice assistant, Siri? 24.

[6] Saggion, H. (2006, May). Multilingual multidocument summarization tools and evaluation. In *LREC* (pp. 1312–1317).

[7] Wong, Y. W., & Mooney, R. J. (2006, June). Learning for semantic parsing with statistical machine translation. In *Proceedings of the Main Conference on Human Language Technology Conference of the North American Chapter of the Association of Computational Linguistics* (pp. 439–446). Association for Computational Linguistics.

[8] Patil, S., & Davies, P. (2014). Use of Google Translate in medical communication: Evaluation of accuracy. *BMJ*, *349*, g7392.

[9] Eckert, M., Bry, F., Brodt, S., Poppe, O., & Hausmann, S. (2011). A CEP babelfish: Languages for complex event processing and querying surveyed. In *Reasoning in event-based distributed systems* (pp. 47–70). Springer, Berlin, Heidelberg.

[10] Aggarwal, C. C., & Zhai, C. (2012). A survey of text classification algorithms. In *Mining text data* (pp. 163–222). Springer, Boston, MA.

[11] Chorowski, J. K., Bahdanau, D., Serdyuk, D., Cho, K., & Bengio, Y. (2015). Attention-based models for speech recognition. In *Advances in neural information processing systems* (pp. 577–585).

[12] Sanjay Agrwal, S. C. D. (2002). DBXplorer: A system for keyword base search over relational database. In *Proceedings of 18th International Conference on Data Engineering*. IEEE.

[13] Chen, P. P. S. (1976). The entity–relationship model – toward a unified view of data. *ACM Transactions on Database Systems (TODS)*, *1*(1), 9–36.

[14] Tseng, F. S., Chen, A. L., & Yang, W. P. (1992). On mapping natural language constructs into relational algebra through ER representation. *Data and Knowledge Engineering*, *9*(1), 97–118.

[15] Queralt, A., & Teniente, E. (2006, November). Reasoning on UML class diagrams with OCL constraints. In *International Conference on Conceptual Modeling* (pp. 497–512). Springer, Berlin, Heidelberg.

[16] Grosz, B. J., Appelt, D. E., Martin, P. A., & Pereira, F. C. (1987). TEAM: An experiment in the design of transportable natural-language interfaces. *Artificial Intelligence*, *32*(2), 173–243.

[17] Owei, V., Rhee, H. S., & Navathe, S. (1997). Natural language query filtration in the conceptual query language. In *Proceedings of the Thirtieth Hawaii International Conference on System Sciences* (Vol. 3, pp. 539–549). IEEE.

[18] Nguyen, D. T., Hoang, T. D., & Pham, S. B. (2002). A Vietnamese natural language interface to database. In *2012 IEEE Sixth International Conference on Semantic Computing* (pp. 130–133). IEEE, China.

[19] Miller, G. A. (1995). WordNet: A lexical database for English. *Communications of the ACM*, *38*(11), 39–41.

[20] Sleator, D. D., & Temperley, D. (1995). Parsing English with a link grammar. *arXiv preprint cmp-lg/9508004*.

[21] Stanford CoreNLP. (2014). Stanford CoreNLP – Natural language software. Last accessed December 23, 2019. Retrieved from: https://stanfordnlp.github.io/CoreNLP/

[22] Johnstone, B. (2018). *Discourse analysis*. John Wiley & Sons.

[23] Leech, G. N. (2016). *Principles of pragmatics*. Routledge.

[24] Giordani, A., & Moschitti, A. (2009, June). Semantic mapping between natural language questions and SQL queries via syntactic pairing. In *International Conference on Application of Natural Language to Information Systems* (pp. 207–221). Springer, Berlin, Heidelberg.

[25] Zhang, J., Scardamalia, M., Reeve, R., & Messina, R. (2009). Designs for collective cognitive responsibility in knowledge-building communities. *The Journal of the Learning Sciences*, *18*(1), 7–44.

[26] Gallè, F., Mancusi, C., Di Onofrio, V., Visciano, A., Alfano, V., Mastronuzzi, R., … & Liguori, G. (2011). Awareness of health risks related to body art practices among youth in Naples, Italy: A descriptive convenience sample study. *BMC Public Health*, *11*(1), 625.

[27] Safari, L., & Patrick, J. D. (2019). An enhancement on Clinical Data Analytics Language (CliniDAL) by integration of free text concept search. *Journal of Intelligent Information Systems*, *52*(1), 33–55.

[28] Kando, N. (1999, November). Text structure analysis as a tool to make retrieved documents usable. In *Proceedings of the 4th International Workshop on Information Retrieval with Asian Languages* (pp. 126–135).

[29] Zhang, M., Zhang, J., & Su, J. (2006, June). Exploring syntactic features for table extraction using a convolution tree kernel. In *Proceedings of the Main Conference on Human Language Technology Conference of the North American Chapter of the Association of Computational Linguistics* (pp. 288–295). Association for Computational Linguistics.

[30] Sagar, R. (2020, June 3). OpenAI releases GPT-3, the largest model so far. *Analytics India Magazine*. Retrieved October 14, 2020.

[31] Chalmers, David (2020, July 30). "GPT-3 and General Intelligence". *Daily Nous*. Last accessed October 14, 2020. Retrieved from: https://dailynous.com/2020/07/30/philosophers-gpt-3/#chalmers

[32] OpenAI, Discovering and enacting the path to safe artificial general intelligence, 2020. Last accessed October 13, 2020. Retrieved from: https://openai.com/

[33] Bybee, J. L. (1985). *Morphology: A study of the relation between meaning and form* (Vol. *9*). John Benjamins Publishing.

[34] Smith, N. V. (1973). *The acquisition of phonology: A case study.* Cambridge University Press.

[35] Stetson, R. H. (2014). *Motor phonetics: A study of speech movements in action.* Springer.

[36] Zhang, D., & Lee, W. S. (2003, July). Question classification using support vector machines. In *Proceedings of the 26th Annual International ACM SIGIR Conference on Research and Development in Information Retrieval* (pp. 26–32). ACM.

[37] Iftikhar, A., Iftikhar, E., & Mehmood, M. K. (2016, August). Domain specific query generation from natural language text. In *2016 Sixth International Conference on Innovative Computing Technology (INTECH)* (pp. 502–506). IEEE.

[38] Kumar, R., & Dua, M. (2014, April). Translating controlled natural language query into SQL query using pattern matching technique. In *International Conference for Convergence for Technology–2014* (pp. 1–5). IEEE.

[39] Boyd-Graber, J., Fellbaum, C., Osherson, D., & Schapire, R. (2006, January). Adding dense, weighted connections to WordNet. In *Proceedings of the Third International WordNet Conference* (pp. 29–36).

[40] NLTK 3.4.5 Documentation. (2019). Natural language toolkit. Last accessed December 23, 2019. Retrieved from: http://www.nltk.org/

[41] Rish, I. (2001, August). An empirical study of the naive Bayes classifier. In *IJCAI 2001 Workshop on Empirical Methods in Artificial Intelligence* (Vol. 3, No. 22, pp. 41–46).

[42] Safari, L., & Patrick, J. D. (2018). Complex analyses on clinical information systems using restricted natural language querying to resolve time-event dependencies. *Journal of Biomedical Informatics, 82,* 13–30.

[43] Ganti, V., He, Y., & Xin, D. (2010). Keyword++: A framework to improve keyword search over entity databases. *Proceedings of the VLDB Endowment, 3*(1–2), 711–722.

[44] Woods, W. A. (1981). *Procedural semantics as a theory of meaning.* Bolt Beranek and Newman Inc.

[45] Kaur, S., & Bali, R. S. (2012). SQL generation and execution from natural language processing. *International Journal of Computing & Business Research,* 2229–6166.

[46] Bhadgale Anil, M., Gavas Sanhita, R., Pati Meghana, M., & Pinki, R. (2013). Natural language to SQL conversion system. *International Journal of Computer Science Engineering and Information Technology Research (IJCSEITR), 3*(2), 161–166, ISSN 2249-6831.

[47] Popescu, A. M., Etzioni, O., & Kautz, H. (2003, January). Towards a theory of natural language interfaces to databases. In *Proceedings of the 8th International Conference on Intelligent User Interfaces* (pp. 149–157). ACM.

[48] Parlikar, A., Shrivastava, N., Khullar, V., & Sanyal, S. (2005). NQML: Natural query markup language. In *2005 International Conference on Natural Language Processing and Knowledge Engineering* (pp. 184–188). IEEE.

[49] Peng, Z., Zhang, J., Qin, L., Wang, S., Yu, J. X., & Ding, B. (2006, September). NUITS: A novel user interface for efficient keyword search over databases. In *Proceedings of the 32nd International Conference on Very Large Data Bases* (pp. 1143–1146). VLDB Endowment.

[50] Feijs, L. M. G. (2000). Natural language and message sequence chart representation of use cases. *Information and Software Technology, 42*(9), 633–647.

[51] Karande, N. D., & Patil, G. A. (2009). Natural language database interface for selection of data using grammar and parsing. *World Academy of Science, Engineering and Technology, 3*, 11–26.

[52] El-Mouadib, F. A., Zubi, Z. S., & Almagrous, A. A. (2009). Generic interactive natural language interface to databases (GINLIDB). *International Journal of Computers, 3*(3).

[53] Li, H., & Shi, Y. (2010, February). A wordnet-based natural language interface to relational databases. In *2010 The 2nd International Conference on Computer and Automation Engineering (ICCAE)* (Vol. 1, pp. 514–518). IEEE.

[54] Enikuomehin, A. O., & Okwufulueze, D. O. (2012). An algorithm for solving natural language query execution problems on relational databases. *International Journal of Advanced Computer Science and Applications, 3*(10), 180–182.

[55] Chen, P. P. S. (1983). English sentence structure and entity-relationship diagrams. *Information Sciences, 29*(2–3), 127–149.

[56] QUEST/a natural language interface to relational databases (2018). In *Proceedings of the Eleventh International Conference on Language Resources and Evaluation (LREC)*.

[57] Desai, B., & Stratica, N. (2004). Schema-based natural language semantic mapping. In *Proceedings of the 9th International Conference on Applications of Natural Language to Information Systems*.

[58] Becker, T. (2002, May). Practical, template–based natural language generation with tag. In *Proceedings of the Sixth International Workshop on Tree Adjoining Grammar and Related Frameworks (TAG+ 6)* (pp. 80–83).

[59] Androutsopoulos, I., Ritchie, G. D., & Thanisch, P. (1995). Natural language interfaces to databases–an introduction. *Natural Language Engineering, 1*(1), 29–81.

[60] Patrick, J., & Li, M. (2010). High accuracy information extraction of medication information from clinical notes: 2009 i2b2 medication extraction challenge. *Journal of the American Medical Informatics Association, 17*(5), 524–527.

[61] Chaudhari, P. (2013). Natural language statement to SQL query translator. *International Journal of Computer Applications, 82*(5), 18–22.

[62] Gao, K., Mei, G., Piccialli, F., Cuomo, S., Tu, J., & Huo, Z. (2020). Julia language in machine learning/algorithms, applications, and open issues. *Computer Science Review, 37*, 100254.

[63] Hasan, R., & Gandon, F. (2014, August). A machine learning approach to sparql query performance prediction. In *2014 IEEE/WIC/ACM International Joint Conferences on Web Intelligence (WI) and Intelligent Agent Technologies (IAT)* (Vol. 1, pp. 266–273). IEEE.

[64] Wang, H., Ma, C., & Zhou, L. (2009, December). A brief review of machine learning and its application. In *2009 International Conference on Information Engineering and Computer Science* (pp. 1–4). IEEE.

[65] Boyan, J., Freitag, D., & Joachims, T. (1996, August). A machine learning architecture for optimizing web search engines. In *AAAI Workshop on Internet Based Information Systems* (pp. 1–8).

[66] Chen, H., Shankaranarayanan, G., She, L., & Iyer, A. (1998). A machine learning approach to inductive query by examples: An experiment using relevance feedback, ID3, genetic algorithms, and simulated annealing. *Journal of the American Society for Information Science, 49*(8), 693–705.

[67] Hazlehurst, B. L., Burke, S. M., & Nybakken, K. E. (1999). Intelligent query system for automatically indexing information in a database and automatically categorizing users. U.S. Patent No. 5,974,412. Washington, DC: U.S. Patent and Trademark Office. WebMD Inc.

[68] Hofmann, T. (2001). Unsupervised learning by probabilistic latent semantic analysis. *Machine Learning, 42*(1–2), 177–196.

[69] Popov, B., Kiryakov, A., Ognyanoff, D., Manov, D., & Kirilov, A. (2004). KIM – A semantic platform for information extraction and retrieval. *Journal of Natural Language Engineering, 10*(3–4), 375–392.

[70] Alexander, R., Rukshan, P., & Mahesan, S. (2013). Natural language web interface for database (NLWIDB). *arXiv preprint arXiv:1308.3830.*

[71] Zhou, Q., Wang, C., Xiong, M., Wang, H., & Yu, Y. (2007). SPARK: Adapting keyword query to semantic search. In *The semantic web* (pp. 694–707). Springer, Berlin, Heidelberg.

[72] Bergamaschi, S., Domnori, E., Guerra, F., Orsini, M., Lado, R. T., & Velegrakis, Y. (2010). Keymantic: Semantic keyword-based searching in data integration systems. *Proceedings of the VLDB Endowment, 3*(1–2), 1637–1640.

[73] Strötgen, J., & Gertz, M. (2010, July). HeidelTime: High quality rule-based extraction and normalization of temporal expressions. In *Proceedings of the 5th International Workshop on Semantic Evaluation* (pp. 321–324). Association for Computational Linguistics.

[74] Sohn, S., Wagholikar, K. B., Li, D., Jonnalagadda, S. R., Tao, C., Komandur Elayavilli, R., & Liu, H. (2013). Comprehensive temporal information detection from clinical text: Medical events, time, and TLINK identification. *Journal of the American Medical Informatics Association, 20*(5), 836–842.

[75] Zhou, L., Friedman, C., Parsons, S., & Hripcsak, G. (*2005*). System architecture for temporal information extraction, representation and reasoning in clinical narrative reports. In *AMIA Annual Symposium Proceedings* (Vol. 2005, p. 869). American Medical Informatics Association.

[76] Giordani, A., & Moschitti, A. (2012, June). Generating SQL queries using natural language syntactic dependencies and metadata. In *International Conference on Application of Natural Language to Information Systems* (pp. 164–170). Springer, Berlin, Heidelberg.

[77] Giordani, A., & Moschitti, A. (2010, May). Corpora for Automatically Learning to Map Natural Language Questions into SQL Queries. In *LREC*.

[78] Kate, R. J., & Mooney, R. J. (2006, July). Using string-kernels for learning semantic parsers. In *Proceedings of the 21st International Conference on Computational Linguistics and the 44th Annual Meeting of the Association for Computational Linguistics* (pp. 913–920). Association for Computational Linguistics.

[79] Tseng, F. S., & Chen, C. L. (2006, September). Extending the UML concepts to transform natural language queries with fuzzy semantics into SQL. *Information and Software Technology, 48*(9), 901–914.

[80] Booch, G. (2005). *The unified modeling language user guide.* Pearson Education India.

[81] Oestereich, B. (2002). *Developing software with UML: Object-oriented analysis and design in practice.* Pearson Education.

[82] Higa, K., & Owei, V. (1991, January). A data model driven database query tool. In *Proceedings of the Twenty-Fourth Annual Hawaii International Conference on System Sciences* (Vol. 3, pp. 53–62). IEEE.

[83] Muller, R. J. (1999). *Database design for smarties: Using UML for data modeling.* Morgan Kaufmann.

[84] Winston, P. H. (1992). *Artificial intelligence.* Addison-Wesley Longman Publishing Co., Inc.

[85] Schmucker, K. J., & Zadeh, L. A. (1984). *Fuzzy sets natural language computations and risk analysis.* Rockville Md/ Computer Science Press.

[86] Zadeh, L. A. (1975). The concept of a linguistic variable and its application to approximate reasoning-I. *Information Sciences, 8,* 199–249.

[87] Isoda, S. (2001). Object-oriented real-world modeling revisited. *Journal of Systems and Software, 59*(2), 153–162.

[88] Moreno, A. M., & Van De Riet, R. P. (1997, June). Justification of the equivalence between linguistic and conceptual patterns for the object model. In *Proceedings of the International Workshop on Applications of Natural Language to Information Systems,* Vancouver.

[89] Métais, E. (2002). Enhancing information systems management with natural language processing techniques. *Data & Knowledge Engineering, 41*(2–3), 247–272.

[90] Yager, R. R., Reformat, M. Z., & To, N. D. (2019). Drawing on the iPad to input fuzzy sets with an application to linguistic data science. *Information Sciences, 479,* 277–291.

[91] Owei, V., Navathe, S. B., & Rhee, H. S. (2002). An abbreviated concept-based query language and its exploratory evaluation. *Journal of Systems and Software, 63*(1), 45–67.

[92] Hoang, D. T., Le Nguyen, M., & Pham, S. B. (2015, October). L2S: Transforming natural language questions into SQL queries. In *2015 Seventh International Conference on Knowledge and Systems Engineering (KSE)* (pp. 85–90). IEEE.

[93] Costa, P. D., Almeida, J. P. A., Pires, L. F., & van Sinderen, M. (2008, November). Evaluation of a rule-based approach for context-aware services. In *IEEE GLOBECOM 2008 – 2008 IEEE Global Telecommunications Conference* (pp. 1–5). IEEE.

[94] Garcia, K. K., Lumain, M. A., Wong, J. A., Yap, J. G., & Cheng, C. (2008, November). Natural language database interface for the community based monitoring system. In *Proceedings of the 22nd Pacific Asia Conference on Language, Information and Computation* (pp. 384–390).

[95] International Conference on Applications of Natural Language to Information Systems (13th: 2008: London, England). (2008). Natural Language and Information

Systems 13th International Conference on Applications of Natural Language to Information Systems, NLDB 2008 London, UK, June 24–27, 2008, Proceedings. Berlin, Heidelberg: Springer Berlin Heidelberg: Imprint: Springer.

[96] Waltz, D. L. (1978). An English language question answering system for a large relational database. *Communications of the ACM, 21*(7), 526–539.

[97] Nguyen, D. B., Hoang, S. H., Pham, S. B., & Nguyen, T. P. (2010, March). Named entity recognition for Vietnamese. In *Asian Conference on Intelligent Information and Database Systems* (pp. 205–214). Springer, Berlin, Heidelberg.

[98] Cunningham, H., Maynard, D., Bontcheva, K., & Tablan, V. (2002, July). GATE: An architecture for development of robust HLT applications. In *Proceedings of the 40th Annual Meeting on Association for Computational Linguistics* (pp. 168–175). Association for Computational Linguistics.

[99] Sathick, K. J., & Jaya, A. (2015). Natural language to SQL generation for semantic knowledge extraction in social web sources. *Indian Journal of Science and Technology, 8*(1), 1–10.

[100] Satav, A. G., Ausekar, A. B., Bihani, R. M., & Shaikh, A. (2014). A proposed natural language query processing system. *International Journal of Science and Applied Information Technology, 3*(2). http://warse.org/pdfs/2014/ijsait01322014.pdf

[101] Kaur, G. (2014). Usage of regular expressions in NLP. *International Journal of Research in Engineering and Technology IJERT, 3*(1), 7.

[102] Agrawal, A. J., & Kakde, O. G. (2013). Semantic analysis of natural language queries using domain ontology for information access from database. *International Journal of Intelligent Systems and Applications, 5*(12), 81.

[103] Deshpande, A. K., & Devale, P. R. (2012). Natural language query processing using probabilistic context free grammar. *International Journal of Advances in Engineering & Technology, 3*(2), 568.

[104] Tamrakar, A., & Dubey, D. (2012). Query Optimisation using Natural Language Processing 1.

[105] Michael, G. (2012). *A Survey of Natural Language Processing Techniques for the Simplification of User Interaction with Relational Database Management Systems*, California Polytechnic State University, San Luis Obispo.

[106] Rao, G., Agarwal, C., Chaudhry, S., Kulkarni, N., & Patil, D. S. (2010). Natural language query processing using semantic grammar. *International Journal on Computer Science and Engineering, 2*(2), 219–223.

[107] Ott, N. (1992). Aspects of the automatic generation of SQL statements in a natural language query interface. *Information Systems, 17*(2), 147–159.

[108] Petrick, S. R. (1984). Natural language database query systems. Technical Report, RC 10508, IBM Thomas J. Watson Research Laboratory.

[109] Lehmann, H., Ott, N., & Zoeppritz, M. (1985). A multilingual interface to databases. *IEEE Database Engineering, 8*(3), 7–13.

[110] Wong, Y. W., & Mooney, R. (2007, June). Learning synchronous grammars for semantic parsing with lambda calculus. In *Proceedings of the 45th Annual Meeting of the Association of Computational Linguistics* (pp. 960–967).

[111] Ge, R., & Mooney, R. (2005, June). A statistical semantic parser that integrates syntax and semantics. In *Proceedings of the Ninth Conference on Computational Natural Language Learning (CoNLL – 2005)* (pp. 9–16).

[112] Minock, M., Olofsson, P., & Näslund, A. (2008, June). Towards building robust natural language interfaces to databases. In *International Conference on Application of Natural Language to Information Systems* (pp. 187–198). Springer, Berlin, Heidelberg.

[113] Zettlemoyer, L. S., & Collins, M. (2012). Learning to map sentences to logical form: Structured classification with probabilistic categorial grammars. *arXiv preprint arXiv:1207.1420*.

[114] Tang, L. R., & Mooney, R. J. (2001, September). Using multiple clause constructors in inductive logic programming for semantic parsing. In *European Conference on Machine Learning* (pp. 466–477). Springer, Berlin, Heidelberg.

[115] Giordani, A., & Moschitti, A. (2012, December). Translating questions to SQL queries with generative parsers discriminatively reranked. In *Proceedings of COLING 2012: Posters* (pp. 401–410).

[116] De Marneffe, M. C., MacCartney, B., & Manning, C. D. (2006, May). Generating typed dependency parses from phrase structure parses. In *LREC* (Vol. 6, pp. 449–454).

[117] Joachims, T. (1998). Making large-scale SVM learning practical (No. 1998, 28). Technical Report.

[118] Moschitti, A. (2006, September). Efficient convolution kernels for dependency and constituent syntactic trees. In *European Conference on Machine Learning* (pp. 318–329). Springer, Berlin, Heidelberg.

[119] Giordani, A., & Moschitti, A. (2009, September). Syntactic structural kernels for natural language interfaces to databases. In *Joint European Conference on Machine Learning and Knowledge Discovery in Databases* (pp. 391–406). Springer, Berlin, Heidelberg.

[120] Lu, W., Ng, H. T., Lee, W. S., & Zettlemoyer, L. (2008, October). A generative model for parsing natural language to meaning representations. In *Proceedings of the 2008 Conference on Empirical Methods in Natural Language Processing* (pp. 783–792).

[121] Liang, P., Jordan, M. I., & Klein, D. (2013). Learning dependency-based compositional semantics. *Computational Linguistics, 39*(2), 389–446.

[122] Clarke, J., Goldwasser, D., Chang, M. W., & Roth, D. (2010, July). Driving semantic parsing from the world's response. In *Proceedings of the Fourteenth Conference on Computational Natural Language Learning* (pp. 18–27). Association for Computational Linguistics.

[123] Kwiatkowski, T., Zettlemoyer, L., Goldwater, S., & Steedman, M. (2010, October). Inducing probabilistic CCG grammars from logical form with higher-order unification. In *Proceedings of the 2010 Conference on Empirical Methods in Natural Language Processing* (pp. 1223–1233). Association for Computational Linguistics.

[124] Xu, X., Liu, C., & Song, D. (2017). SQLnet: Generating structured queries from natural language without reinforcement learning. *arXiv preprint arXiv: 1711.04436*.

[125] Thompson, B. H., & Thompson, F. B. (1985). ASK is transportable in half a dozen ways. *ACM Transactions on Information Systems, 3*(2), 185–203.

[126] Kudo, T., Suzuki, J., & Isozaki, H. (2005, June). Boosting-based parse reranking with subtree features. In *Proceedings of the 43rd Annual Meeting on Association for Computational Linguistics* (pp. 189–196). Association for Computational Linguistics.

[127] Toutanova, K., Markova, P., & Manning, C. (2004). The leaf path projection view of parse trees: Exploring string kernels for HPSG parse selection. In *Proceedings of the 2004 Conference on Empirical Methods in Natural Language Processing* (pp. 166–173).

[128] Kazama, J. I., & Torisawa, K. (2005, October). Speeding up training with tree kernels for node table labeling. In *Proceedings of the Conference on Human Language Technology and Empirical Methods in Natural Language Processing* (pp. 137–144). Association for Computational Linguistics.

[129] Gaikwad Mahesh, P. (2013). Natural language interface to database. *International Journal of Engineering and Innovative Technology (IJEIT)*, 2(8), 3–5.

[130] Papalexakis, E., Faloutsos, C., & Sidiropoulos, N. D. (2012). ParCube: Sparse parallelizable tensor decompositions. In *Joint European Conference on Machine Learning and Knowledge Discovery in Databases* (pp. 521–536). Springer, Berlin, Heidelberg.

[131] Safari, L., & Patrick, J. D. (2014). Restricted natural language based querying of clinical databases. *Journal of Biomedical Informatics*, 52, 338–353.

[132] Chandra, Y., & Mihalcea, R. (2006). Natural language interfaces to databases, University of North Texas (Doctoral dissertation, Thesis (MS)).

[133] Shen, L., Sarkar, A., & Joshi, A. K. (2003, July). Using LTAG based features in parse reranking. In *Proceedings of the 2003 Conference on Empirical Methods in Natural Language Processing* (pp. 89–96). Association for Computational Linguistics.

[134] Collins, M., & Duffy, N. (2002, July). New ranking algorithms for parsing and tagging: Kernels over discrete structures, and the voted perceptron. In *Proceedings of the 40th Annual Meeting on Association for Computational Linguistics* (pp. 263–270). Association for Computational Linguistics.

[135] Kudo, T., & Matsumoto, Y. (2003, July). Fast methods for kernel-based text analysis. In *Proceedings of the 41st Annual Meeting on Association for Computational Linguistics-Volume 1* (pp. 24–31). Association for Computational Linguistics.

[136] Cumby, C. M., & Roth, D. (2003). On kernel methods for relational learning. In *Proceedings of the 20th International Conference on Machine Learning (ICML-03)* (pp. 107–114).

[137] Culotta, A., & Sorensen, J. (2004, July). Dependency tree kernels for table extraction. In *Proceedings of the 42nd Annual Meeting on Association for Computational Linguistics* (p. 423). Association for Computational Linguistics.

[138] Ghosal, D., Waghmare, T., Satam, S., & Hajirnis, C. (2016). SQL query formation using natural language processing (NLP). *International Journal of Advanced Research in Computer and Communication Engineering*, 5, 3.

[139] Zhang, J., Tang, J., Ma, C., Tong, H., Jing, Y., & Li, J. (2015, August). Panther: Fast top-k similarity search on large networks. In *Proceedings of the 21st ACM SIGKDD International Conference on Knowledge Discovery and Data Mining* (pp. 1445–1454).

[140] Ghosh, P. K., Dey, S., & Sengupta, S. (2014). Automatic SQL query formation from natural language query. *International Journal of Computer Applications, 975*, 8887.

[141] Choudhary, N., & Gore, S. (2015, September). Impact of intellisense on the accuracy of natural language interface to database. In *2015 4th International Conference on Reliability, Infocom Technologies and Optimization (ICRITO) (Trends and Future Directions)* (pp. 1–5). IEEE.

[142] Willett, P. (2006). The Porter stemming algorithm: Then and now. Program.

[143] Yu, T., Zhang, R., Yang, K., Yasunaga, M., Wang, D., Li, Z., ... & Zhang, Z. (2018). Spider: A large-scale human-labeled dataset for complex and cross-domain semantic parsing and text-to-SQL task. *arXiv preprint arXiv:1809.08887*.

[144] Nelken, R., & Francez, N. (2000, July). Querying temporal databases using controlled natural language. In *Proceedings of the 18th Conference on Computational Linguistics – Volume 2* (pp. 1076–1080). Association for Computational Linguistics.

[145] Naumann, F. (2014). Data profiling revisited. *ACM SIGMOD Record, 42*(4), 40–49.

[146] Singh, G., & Solanki, A. (2016). An algorithm to transform natural language into SQL queries for relational databases. *Selforganizology, 3*(3), 100–116.

[147] Zadeh, L. A. (1965). Fuzzy sets. *Information and Control, 8*(3), 338–353.

[148] Zadeh, L. A. (1978). PRUF – A meaning representation language for natural languages. *International Journal of Man-Machine Studies, 10*(4), 395–460.

[149] Dalrymple, M., Shieber, S. M., & Pereira, F. C. (1991). Ellipsis and higher-order unification. *Linguistics and Philosophy, 14*(4), 399–452.

[150] Tanaka, H., & Guo, P. (1999). Portfolio selection based on upper and lower exponential possibility distributions. *European Journal of Operational Research, 114*(1), 115–126.

[151] Kang, I.-S., Bae, J.-H., & Lee, J.-H. Database semantics representation for natural language access. In *Proceedings of the First International Symposium on Cyber Worlds (CW '02)*, ISBN:0-7695-1862-1.

[152] De Marneffe, M. C., & Manning, C. D. (2008). Stanford typed dependencies manual (pp. 338–345). Technical Report, Stanford University.

[153] Zeng, J., Lin, X. V., Xiong, C., Socher, R., Lyu, M. R., King, I., & Hoi, S. C. H. (2020). Photon: A robust cross-domain text-to-SQL system. In *Proceedings of the 58th Annual Meeting of the Association for Computational Linguistics: System Demonstrations* (pp. 204–214). ACL.

[154] Poole, D. L., & Mackworth, A. K. (2010). *Artificial intelligence: Foundations of computational agents*. Cambridge University Press.

[155] Warren, D. H., Pereira, L. M., & Pereira, F. (1977). Prolog – the language and its implementation compared with Lisp. *ACM SIGPLAN Notices, 12*(8), 109–115.

[156] Wang, P., Shi, T., & Reddy, C. K. (2019). A translate-edit model for natural language question to SQL query generation on multi-relational healthcare Data. *arXiv preprint arXiv:1908.01839*.

[157] Yao, K., & Zweig, G. (2015). Sequence-to-sequence neural net models for grapheme-to-phoneme conversion. *arXiv preprint arXiv:1506.00196*.

[158] Zhang, Z., & Sabuncu, M. (2018). Generalized cross-entropy loss for training deep neural networks with noisy labels. In *Advances in Neural Information Processing Systems* (pp. 8778–8788).

[159] Seo, M., Kembhavi, A., Farhadi, A., & Hajishirzi, H. (2016). Bidirectional attention flow for machine comprehension. *arXiv preprint arXiv:1611.01603*.

[160] Ke, N. R., Goyal, A. G. A. P., Bilaniuk, O., Binas, J., Mozer, M. C., Pal, C., & Bengio, Y. (2018). Sparse attentive backtracking: Temporal credit assignment through reminding. In *Advances in neural information processing systems* (pp. 7640–7651).

[161] Michaelsen, S. M., Dannenbaum, R., & Levin, M. F. (2006). Task-specific training with trunk restraint on arm recovery in stroke: Randomized control trial. *Stroke*, *37*(1), 186–192.

[162] Löb, M. H. (1976). Embedding first order predicate logic in fragments of intuitionistic logic. *The Journal of Symbolic Logic*, *41*(4), 705–718.

[163] Yu, B., Lin, X., & Wu, Y. (1991). The tree representation of the graph used in binary image processing. *Information Processing Letters*, *37*(1), 55–59.

[164] Python. (2018, June). Python 3.7.0 Home Page. Last accessed December 23, 2019. Retrieved from: https://www.python.org/downloads/release/python-370/

[165] MySQL Community Downloads. (2019). MySQL Community Server 8.0.18 Home Page. Last accessed December 23, 2019. Retrieved from: https://dev. mysql.com/downloads/mysql/

[166] MySQLdb. (2012). Welcome to MySQLdb's documentation! Last accessed December 23, 2019. Retrieved from: https://mysqldb.readthedocs.io/en/latest/

[167] TextBlob. (2015). TextBlob: Simplified Text Processing. Last accessed December 23, 2019. Retrieved from: https://textblob.readthedocs.io/en/dev/

[168] JetBrains. (2019). IDE PyCharm C Home Page. Last accessed December 23, 2019. Retrieved from: https://www.jetbrains.com/pycharm/

[169] XQuartz. (2016, October). XQuartz 2.7.11 Home Page. Last accessed December 23, 2019. Retrieved from: https://www.xquartz.org/index.html

[170] Apple Developer. (2019). Xcode 11 Home Page. Last accessed December 23, 2019. Retrieved from: https://developer.apple.com/xcode/

[171] MySQL. (2019). MySQL Workbench Home Page. Last accessed December 23, 2019. Retrieved from: https://www.mysql.com/products/workbench/

[172] Kaggle. (2017). Zomato Restaurants Data. Last accessed December 23, 2019. Retrieved from: https://www.kaggle.com/shrutimehta/zomato-restaurants-data

[173] GitHub. (2017). WikiSQL RDB – A large annotated semantic parsing corpus for developing natural language interfaces. Last accessed December 23, 2019. Retrieved from: https://github.com/salesforce/WikiSQL

[174] Zhong, V., Xiong, C., & Socher, R. (2017). Seq2SQL: Generating structured queries from natural language using reinforcement learning. *arXiv preprint arXiv:1709.00103*.

[175] Zhong, V., Xiong, C., & Socher, R. (2017). Seq2SQL: Generating structured queries from natural language using reinforcement learning. *arXiv preprint arXiv:1709.00103*.

[176] Streiner, D. L., & Cairney, J. (2007). What's under the ROC? An introduction to receiver operating characteristics curves. *The Canadian Journal of Psychiatry*, *52*(2), 121–128.

[177] Original implementation code extended from "Couderc, B., & Ferrero, J. (2015, June). fr2SQL: Interrogation de bases de données en français".

[178] Yu, T., Li, Z., Zhang, Z., Zhang, R., & Radev, D. (2018). TypeSQL: Knowledge-based type-aware neural text-to-SQL generation. *arXiv preprint arXiv:1804.09769.*

[179] Hwang, W., Yim, J., Park, S., & Seo, M. (2019). A comprehensive exploration on wikisql with table-aware word contextualization. *arXiv preprint arXiv: 1902.01069.*

[180] He, P., Mao, Y., Chakrabarti, K., & Chen, W. (2019). X-SQL: Reinforce schema representation with context. *arXiv preprint arXiv:1908.08113.*

[181] Yavuz, S., Gur, I., Su, Y., & Yan, X. (2018). What it takes to achieve 100% condition accuracy on WikiSQL. In *Proceedings of the 2018 Conference on Empirical Methods in Natural Language Processing* (pp. 1702–1711).

[182] Gur, I., Yavuz, S., Su, Y., & Yan, X. (2018, July). DialSQL: Dialogue based structured query generation. In *Proceedings of the 56th Annual Meeting of the Association for Computational Linguistics* (Vol. 1: Long Papers pp. 1339–1349).

[183] Zhekova, M., & Totkov, G. (2021). Question patterns for natural language translation in SQL queries. *International Journal on Information Technologies & Security, 13*(2), 43–54.

[184] Brunner, U., & Stockinger, K. (2021, April). Valuenet: A natural language-to-sql system that learns from database information. In *2021 IEEE 37th International Conference on Data Engineering (ICDE)* (pp. 2177–2182). IEEE.

[185] Xu, X., Liu, C., & Song, D. (2017). SQLnet: Generating structured queries from natural language without reinforcement learning. *arXiv preprint arXiv: 1711.04436.*

[186] Talreja, R., & Whitt, W. (2008). Fluid models for overloaded multiclass many-server queueing systems with first-come, first-served routing. *Management Science, 54*(8), 1513–1527.

[187] Tosirisuk, P., & Chandra, J. (1990). Multiple finite source queueing model with dynamic priority scheduling. *Naval Research Logistics (NRL), 37*(3), 365–381.

[188] Carbonell, J. R., Ward, J. L., & Senders, J. W. (1968). A queueing model of visual sampling experimental validation. *IEEE Transactions on Man-Machine Systems, 9*(3), 82–87.

[189] Hoi, S. Y., Ismail, N., Ong, L. C., & Kang, J. (2010). Determining nurse staffing needs: The workload intensity measurement system. *Journal of Nursing Management, 18*(1), 44–53.

[190] Robinson, W. N. (2003, September). Monitoring web service requirements. In *Proceedings 11th IEEE International Requirements Engineering Conference, 2003* (pp. 65–74). IEEE.

[191] Kim, C., & Kameda, H. (1990). Optimal static load balancing of multi-class jobs in a distributed computer system. *IEICE Transactions (1976–1990), 73*(7), 1207–1214.

Index

Pages in *italics* refer to figures and pages in **bold** refer to tables.